ANNALS OF MATHEMATICS STUDIES

Number 29

ANNALS OF MATHEMATICS STUDIES

Edited by Emil Artin and Marston Morse

1. Algebraic Theory of Numbers, *by* HERMANN WEYL
3. Consistency of the Continuum Hypothesis, *by* KURT GÖDEL
6. The Calculi of Lambda-Conversion, *by* ALONZO CHURCH
7. Finite Dimensional Vector Spaces, *by* PAUL R. HALMOS
10. Topics in Topology, *by* SOLOMON LEFSCHETZ
11. Introduction to Nonlinear Mechanics, *by* N. KRYLOFF *and* N. BOGOLIUBOFF
14. Lectures on Differential Equations, *by* SOLOMON LEFSCHETZ
15. Topological Methods in the Theory of Functions of a Complex Variable, *by* MARSTON MORSE
16. Transcendental Numbers, *by* CARL LUDWIG SIEGEL
17. Problème Général de la Stabilité du Mouvement, *by* M. A. LIAPOUNOFF
19. Fourier Transforms, *by* S. BOCHNER *and* K. CHANDRASEKHARAN
20. Contributions to the Theory of Nonlinear Oscillations, Vol. I, *edited by* S. LEFSCHETZ
21. Functional Operators, Vol. I, *by* JOHN VON NEUMANN
22. Functional Operators, Vol. II, *by* JOHN VON NEUMANN
23. Existence Theorems in Partial Differential Equations, *by* DOROTHY L. BERNSTEIN
24. Contributions to the Theory of Games, Vol. I, *edited by* H. W. KUHN *and* A. W. TUCKER
25. Contributions to Fourier Analysis, *by* A. ZYGMUND, W. TRANSUE, M. MORSE, A. P. CALDERON, *and* S. BOCHNER
26. A Theory of Cross-Spaces, *by* ROBERT SCHATTEN
27. Isoperimetric Inequalities in Mathematical Physics, *by* G. POLYA *and* G. SZEGO
28. Contributions to the Theory of Games, Vol. II, *edited by* H. KUHN *and* A. W. TUCKER
29. Contributions to the Theory of Nonlinear Oscillations, Vol. II, *edited by* S. LEFSCHETZ
30. Contributions to the Theory of Riemann Surfaces, *edited by* L. AHLFORS *et al.*

CONTRIBUTIONS TO THE THEORY OF NONLINEAR OSCILLATIONS

VOLUME II

M. L. CARTWRIGHT S. LEFSCHETZ
E. A. CODDINGTON N. LEVINSON
H. F. DeBAGGIS J. McCARTHY
H. L. TURRITTIN

EDITED BY S. LEFSCHETZ

PRINCETON
PRINCETON UNIVERSITY PRESS
1952

The papers in this volume by DeBaggis, Lefschetz, and Turrittin, were prepared under contract with the Office of Naval Research, and equally sponsored by the Office of Air Research. The paper by Coddington and Levinson was prepared under contract with the Office of Naval Research.

Printed in the United States of America

PREFACE

This monograph, a sequel to Annals of Mathematics Study No. 20, offers another collection of contributions to differential equations. All but the last deal more or less with oscillatory problems. In Mary L. Cartwright's paper there are given new and more accurate estimates of the period and amplitude of the oscillation of the van der Pol equation

$$\ddot{x} + \mu(x^2 - 1)\dot{x} + x = 0 \tag{1}$$

for μ large.

E. A. Coddington and Norman Levinson examine sufficient conditions for the existence of periodic solutions of

$$\dot{x} = Ax + \mu + \quad (x, t, \mu) \tag{2}$$

where x is an n-vector, A a constant matrix, μ a small parameter and f periodic in t.

DeBaggis gives n.a.s.c. for the structural stability of a system

$$\dot{x} = f(x, y), \; \dot{y} = g(x, y). \tag{3}$$

The paper by Lefschetz consists of two parts. In the first there is given a complete description of the critical points of an analytical system (3). In the second the solutions of the equation of van der Pol in the full phase plane are studied and described.

The title of the paper by John McCarthy tells its own story.

H. L. Turrittin disposes completely of the formal problem of the solution of an equation

$$\varepsilon^h \, . \, \dot{x} = A(t, \varepsilon)x \tag{4}$$

where ε is a parameter, $\dot{x}$ is an n-vector and A an n matrix whose terms are power series in ε. He also shows that the formal solutions are, under certain conditions, actual solutions.

S. Lefschetz

Princeton University
January, 1952

CONTENTS

CONTRIBUTIONS TO THE THEORY OF

NONLINEAR OSCILLATIONS

VOL. II

I. VAN DER POL'S EQUATION FOR RELAXATION OSCILLATIONS

By M. L. Cartwright

§1. Introduction. The equation

$$\ddot{x} - k(1 - x^2)\dot{x} + x = 0 \tag{1}$$

with k large and positive has only one periodic solution, other than $x = 0$, and this is of a type usually described as a relaxation oscillation (as opposed to a sinusoidal oscillation). It was discussed by van der Pol[1] who obtained a graphical solution for $k = 10$ and by le Corbeiller[2] who, using Liénard's method, showed that the period $2T = 2k(3/2 - \log_e 2) + \underline{o}(k)$, and the greatest height $h = 2 + o(1)$ as $k \to \infty$. Other authors[3] have also discussed the equation, in particular Dorodnitsin[4] has obtained an asymptotic formula for T with smaller error terms but his analysis is difficult to follow.

This paper is based on the joint work of Professor J. E. Littlewood and myself, largely on work which was done before that contained in our other published papers on nonlinear differential equations. We shall show that as $k \to \infty$

$$T = k(3/2 - \log_e 2) + \frac{3(\alpha + \beta)}{2k^{1/3}} + \underline{o}\left(\frac{1}{k^{1/3}}\right) \tag{2}$$

$$h = 2 + \frac{\alpha + \beta}{3k^{4/3}} + \underline{o}\left(\frac{1}{k^{4/3}}\right), \tag{3}$$

where α and β are constants determined as follows: The equation

$$\eta_0 \frac{d\eta_0}{d\xi} = 2\xi\eta_0 + 1 \tag{4}$$

has one and only one solution $\eta_0^*(\xi)$ such that $\eta_0^*(\xi) \to 0$ as $\xi \to -\infty$

(1) B. van der Pol, Phil, Mag. 2 (1926), 978-992.
(2) Ph. le Corbeiller, Journal Inst. Elec. Eng., 79 (1936), 361-378.
(3) D. A. Flanders and J. J. Stoker, "The Limit Case of Relaxation Oscillations", in "Studies in Nonlinear Vibration Theory", ed. R. Courant (Inst. for Maths. and Mech., New York University, 1946, typescript), J. Haag, Ann. Ec. Norm. Sup. 60 (1943), 35-111, 61 (1944), 73-117, J. P. LaSalle, Quart. App. Maths., 7 (1949), 1-20.
(4) A. A. Dorodnitsin, "Prik. Mat. i. Mech.", 11 (1947).

(5) $$\alpha = \eta_0^*(0), \beta = \int_0^\infty \frac{d\xi}{\eta_0^*(\xi)} .$$

It has been shown elsewhere[5] that a solution for which $x = x_0$, $\dot{x} = \dot{x}_0$ at $t = 0$ satisfies

(6) $$|x| < M_0, \quad |\dot{x}| < M_0 k,$$

where M_0 is an absolute constant, for $t > t_0\ (M_0, x_0, \dot{x}_0)$, and restricting ourselves to solutions satisfying (6), we show in the preliminary lemmas that with the exception of certain classes the solutions settle down within half a cycle to a form very similar to that of the periodic solution; a more detailed summary is given in §9, and the final results for the periodic solution are given in Theorem 2 in § 12.

§2. The constants α and β are determined by the behaviour of the solution as it crosses $x = 1$ downwards, and we shall give special consideration to the strip

$$|x - 1| \leq \Delta k^{-2/3}$$

where Δ is arbitrarily large. We use δ and ϵ for small positive numbers and M for an absolute constant, that is depending possibly on M_0 but not on Δ, δ, or ϵ; M_0 retains its identity throughout, but M without a suffix is not necessarily the same in each place. Capital letters (other than M and T) refer to points and small letters to the value of x at the corresponding point, so that H is the point at which x attains a maximum h. We also use small letters with dots to denote the value of $\dot{x}$ at the corresponding point.

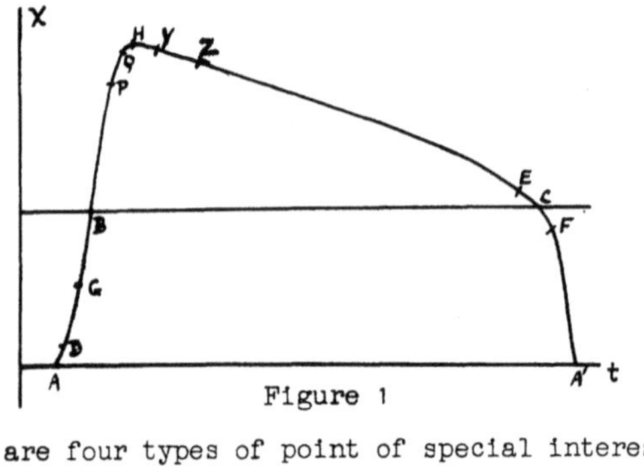

Figure 1

There are four types of point of special interest, A, B, H, C; at A the solution crosses $x = 0$ upwards, at B it crosses $x = 1$ upwards, H is a maximum above $x = 1$ and at C the solution crosses $x = 1$ downwards. It is obvious that there is no minimum above $x = 0$, and that the periodic solution rises above $x = 1$. The equation is unchanged if the signs of both x and $\dot{x}$ are changed so that there is an arc $A' B' H' C'$ below $x = 0$ corresponding to A B H C above $x = 0$, and we deal with the arc $A' B' H' C'$ A by reflecting

(5) See for instance Theorem 1, p. 162, in Forced Oscillations in Nonlinear Systems by M. L. Cartwright in Contributions to the Theory of Nonlinear Oscillations, ed. S. Lefschetz, Annals of Math. Studies, 20, (1950).

the results for A B H C A'. We use t_{ab}, t_{bd}, and so on to denote the time from A to B, B to H and so on.

From what has just been said, it follows that $a = 0$, $\dot{a} > 0$, $b = 1$, $\dot{b} > 0$, $h > 1$, $\dot{h} = 0$, $c = 1$, $\dot{c} < 0$, $a' = 0$, $\dot{a}' < 0$, and it follows from § 1 (6) that $h < M_o$, $\dot{a} < M_o k$, $\dot{b} < M_o k$ and $\dot{c} > - M_o k$, for $t > t_o(M_o, x_o, \dot{x}_o)$.

Besides the equation itself and the simple integrals

$$(1) \qquad x - x_o = \int_{t_o}^{t} \dot{x}(t)dt, \quad \dot{x} - \dot{x}_o = \int_{t_o}^{t} \ddot{x}(t)dt,$$

and

$$(2) \qquad t - t_o = \int_{x_o}^{x} \frac{dx}{\dot{x}},$$

we use the <u>integrated equation</u>

$$(3) \qquad \dot{x} - \dot{x}_o = k\left(x - \frac{x^3}{3} - x_o + \frac{x_o^3}{3}\right) - \int_{t_o}^{t} x\, dt,$$

and the <u>energy equation</u>

$$(4) \qquad \dot{x}^2 - \dot{x}_o^2 = 2k \int_{t_o}^{t} (1-x^2)\, \dot{x}^2\, dt - x^2 + x_o^2,$$

where $x_o = x(t_o)$, $\dot{x}_o = \dot{x}(t_o)$. It should be observed that a change of origin of t leaves the original equation §1 (1) unchanged and so we can shift it at our convenience.

§3. The two following lemmas are remarkàble in that they tell us so much about a solution in the stretch immediately following any maximum for which $1 \leq h < M_o$.

LEMMA 1. A solution starting at H reaches C with $\dot{c}^2 \leq h^2 - 1$.

By §2 (4) we have

$$\dot{c}^2 = 2k \int_H^C (1 - x^2)\, \dot{x}^2\, dt - 1 + h^2 \leq h^2 - 1.$$

LEMMA 2. A solution starting from H reaches any point X in HC with

(i) $\ddot{x} \leq 0$,

(ii) $0 \geq \dot{x} \geq - \dfrac{x}{k(x^2 - 1)}$,

(iii) $\dddot{x} \geq - \dfrac{h^3}{2k(x-1)^2}$

(iv) $t_{xy} \geq k\, \{ \frac{1}{2}(x^2 - y^2) + \log y/x\}$ for $h \geq x > y \geq 1$, in particular $t_{hc} \geq k\, \{ \frac{1}{2}(h^2 - 1) - \log h\}$.

Let $\varphi(x) = (x^2 - 1)/x$ so that $\varphi'(x) = 1 + 1/x^2 > 0$. Hence φ increases with x and since $\dot{x} < 0$ in HC, this means that $1/\varphi$ increases with t. Suppose that $\ddot{z} > 0$ at some point Z in HC, then since $\ddot{h} < 0$ and $\ddot{x}$ is continuous there is a sub-arc XZ of HC such that $\ddot{x} = 0$ and $\ddot{y} > 0$ for $x > y \geq z$.

Hence

$$\dot{x} = \frac{-x}{k(x^2-1)} = -\frac{1}{k\,\varphi(x)},$$

and

$$\dot{y} = -\frac{y + \ddot{y}}{k(y^2-1)} < -\frac{1}{k\,\varphi(y)}, \quad x > y \geq z.$$

But

$$0 > -\frac{1}{k\,\varphi(z)} + \frac{1}{k\,\varphi(x)} > \dot{z} - \dot{x} = \int_X^Z \ddot{y}\,dt > 0,$$

which gives a contradiction, and so $\ddot{x} \leq 0$ for all X in HC.

Part (ii) follows at once from (i) and the original equation, and also if $h > x > y > 1$.

$$t_{xy} = \int_y^x \frac{dx}{|\dot{x}|} > k \int_y^x \left(x - \frac{1}{x}\right) dx = k\{\tfrac{1}{2}(x^2 - y^2) - \log x/y.$$

which gives (iv) and the particular result follows.

Finally

$$\begin{aligned}\dddot{x} &= -k(x^2-1)\ddot{x} - 2kx\dot{x}^2 - \dot{x} \\ &= k(x^2-1)|\ddot{x}| - 2kx\dot{x}^2 + |\dot{x}| \\ &\geq -2kx\dot{x}^2 > -\frac{2x^3}{k(x^2-1)^2} > -\frac{h^3}{2k(x-1)^2},\end{aligned}$$

which is (3).

§4. Lemma 2 consists of one-sided inequalities except in (ii), and it is obvious that nothing better than the left-hand inequality in (ii) can be obtained near H, but in order to obtain a more accurate estimate of t_{hc} we need inequalities of the opposite kind. For this purpose we need to know that h is not too near 1 and we divide the arc HC by points Y, Z and E.

LEMMA 3. Suppose that $h > \frac{3}{2}$ and that HY is an arc on which

(1) $$\dot{x} > -\frac{x - \delta}{k(x^2-1)},$$

where $0 < \delta < h$. Then (i) $\ddot{y} \leq -\delta$; (ii) $t_{hy} < M/(\delta k)$, (iii) $y > h - M/(\delta k^2)$ for $k > k_0(\delta)$.

Since the right hand side of (1) is certainly negative for x sufficiently near h, the interval HY certainly exists. Since the right hand side changes sign at $x = \delta$ and $x = 1$ and $\dot{x} < 0$ in HC, $y \geq 1$. In HY

$$\ddot{x} = k\,|\dot{x}|\,(x^2-1) - x < -\delta$$

so that we have (i). Also

(2) $$\dot{y} = \int_H^Y \ddot{x}\,dt < -\delta t_{hy}.$$

But by Lemma 2 (iv)

$$t_{hy} \geq k\{ \tfrac{1}{2}(h^2 - y^2) \log h/y\} ,$$

and so

(3) $$k\{ \tfrac{1}{2}(h^2 - y^2) - \log h/y\} \leq t_{hy} < |\dot{y}|/\delta \leq \frac{h - \delta}{2k(y-1)\delta}$$

Since $h > \frac{3}{2}$, this gives a contradiction for $y > 1\frac{1}{4}$ unless (iii) holds, and then (ii) follows from (2) and the right hand side of (3).

LEMMA 4. Suppose that HY is the longest arc satisfying the hypotheses of Lemma 3 and that YZ is an arc in which

(4) $$(x^2 - 1)\ddot{x} \leq -3x\dot{x}^2$$

Then

(i) $\ddot{z} \leq -M/k^2$,

(ii) $t_{yz} < (M \log k)/k$,

(iii) $z > h - (M \log k)/k^2$,

provided that $\delta < \delta_0$, and $k > k_0(\delta)$.

In virtue of Lemma 3 (i) and Lemma 2 (ii) (4) obviously holds at Y. By Lemma 2 $\ddot{x} < 0$ in YZ. Differentiating the original equation we have

(5) $$\begin{aligned} \dddot{x} &= k(x^2 - 1)|\ddot{x}| - 2kx\dot{x}^2 + |\dot{x}| \\ &> k(x^2 - 1)|\ddot{x}| - 2kx\dot{x}^2 \\ &> \tfrac{1}{3}k(x^2 - 1)|\ddot{x}| > \tfrac{1}{3}k|\ddot{x}| \end{aligned}$$

for that part of YZ in which $x^2 > 2$. Since $y > h - M/(\delta k^2) \geq \frac{3}{2} - M/(\delta k^2)$, there certainly is such a part for every $\delta > 0$, provided that $k > k_0(\delta)$. It follows that

$$-\log\left|\frac{\ddot{z}}{\ddot{y}}\right| = \int_Y^Z \frac{\dddot{x}\,dt}{|\ddot{x}|} > \tfrac{1}{3}k\,t_{yz}$$

Now $|\ddot{y}| = \delta$, and as long as $|\ddot{z}| < \frac{1}{2}$, we have $|\dot{z}| \geq \frac{1}{2}z/\{k(z^2 - 1)\}$. Hence by (4) above

$$|\ddot{z}| \geq \frac{3z\dot{z}^2}{z^2-1} \geq \frac{M}{k^2} ,$$

provided that $z^2 > 2$. Hence

$$t_{yz} < \frac{M\log(k^2\delta)}{k} < \frac{M\log k}{k} ,$$

provided that $z^2 > 2$, and $k > k_0(\delta)$. But by Lemma 2 (iv) in this time x can only reach a point $z > h - (M \log k)/k^2$, and so $z^2 > 2$ for $k > k_0$, and (ii) and (iii) hold.

LEMMA 5. Suppose that YZ is the longest arc satisfying the hypotheses of Lemma 4, and that E is a point in ZC at which $e \geq 1 + \Delta k^{-2/3}$, then if X is any point in ZE

$$\text{(i)}\quad |\ddot{x}| < \frac{M}{k^2(x-1)^3}\ ,$$

$$\text{(ii)}\quad \dot{x} < \frac{x}{k(x^2-1)}\left(1 - \frac{M}{k^2(x-1)^3}\right),$$

$$\text{(iii)}\quad t_{ze} < k\{\tfrac{1}{2}(z^2 - e^2) - \log z/e\} + \frac{M}{\Delta k^{1/3}} \quad \text{for } \Delta > \Delta_o\ .$$

At Z (i) holds in virtue of Lemma 2 (ii) and (4), and since $\ddot{x} < 0$ it will continue to hold as long as $\dddot{x} > 0$ which is the case at Z by (5). But if $\dddot{x} \leq 0$

$$k(x^2 - 1)\,|\ddot{x}| \leq 2kx\dot{x}^2 - |\dot{x}| < \frac{M}{k(x-1)^2}$$

by Lemma 2 (ii), so that (i) still holds throughout any part of ZE in which $\dddot{x} \leq 0$. But in any other interval $\dddot{x} > 0$ and $\ddot{x}$ is increasing so that we can repeat the argument, and (i) holds throughout ZE.

Part (ii) follows at once, and

$$t_{ze} = \int_E^Z \frac{dx}{|\dot{x}|} \leq k \int_E^Z \left(x - \frac{1}{x}\right)\left(1 - \frac{M}{k^2(x-1)^3}\right)^{-1} dx$$

$$\leq k\{\tfrac{1}{2}(z^2 - e^2) - \log \tfrac{z}{e}\} + \frac{M}{k}\int_E^Z \frac{dx}{(x-1)^2}$$

$$\leq k\{\tfrac{1}{2}(z^2 - e^2) - \log \tfrac{z}{e}\} + \frac{M}{\Delta k^{1/3}}$$

provided that $\Delta > \Delta_o$.

LEMMA 6. If E satisfies the hypotheses of Lemma 5 and $e = 1 + \Delta k^{-2/3}$, then (i) in EC

$$\frac{M}{\Delta k^{1/3}} \leq |\dot{x}| \leq M\,\frac{\Delta^{1/2}}{k^{1/3}}$$

and (ii) $t_{ec} < M\Delta^2/k^{1/3}$, provided that $\Delta > \Delta_o$.

By Lemma 5 for $\Delta > \Delta_o$

$$(6)\qquad |\dot{e}| > \frac{e}{2k(e^2-1)} > \frac{M}{\Delta k^{1/3}}\ .$$

Also from the energy equation

$$(7)\qquad \dot{c}^2 \leq \dot{e}^2 + e^2 - 1 \leq \frac{M}{\Delta^2 k^{2/3}} + \frac{M\Delta}{k^{2/3}} \leq \frac{M\Delta}{k^{2/3}}\ .$$

Since $\ddot{x} < 0$ in EC, (i) follows from (6) and (7).
Then

$$t_{ec} = \int_C^E \frac{dx}{|\dot{x}|} \leq \frac{e-1}{|\dot{e}|} \leq \frac{\Delta}{k^{2/3}}\cdot M\Delta k^{1/3}\ ,$$

which gives (ii).

§5. Before proceeding further we may observe that

$$t_{hc} = t_{hy} + t_{yz} + t_{ze} + t_{ec}$$

(1) $$= \frac{\Psi_1}{k} + \Psi_2 \frac{\log k}{k} + k\{\tfrac{1}{2}(z^2 - e^2) - \log \tfrac{z}{e}\} + \frac{\Psi_3}{\Delta k^{1/3}} + \frac{\Psi_4 \Delta^2}{k^{1/3}}$$

$$= k\{\tfrac{1}{2}(h^2 - 1) - \log h\} - k(e - 1)^2 + \Psi_4 \frac{\Delta^2}{k^{1/3}} + \frac{\Psi_5}{\Delta k^{1/3}},$$

where $|\Psi_s| < M$ for $s = 1, 2, 3, 4, 5$, provided that $\Delta > \Delta_0$ and $k > k_0(\delta, \Delta)$. The two largest error terms $k(e - 1)^2$ and $\Psi_4 \Delta^2 k^{-1/3}$ come from the interval EC which will need special consideration.

§6. We now proceed to study the arc CA', framing our lemmas so as to cover all solutions starting down from C.

LEMMA 7. If a solution starts from C, and if F is a point in CA' at which $f \leq 1 - \Delta k^{-2/3}$, where $\Delta > 0$, then

(i) $|\dot{f}| > \max(\Delta^{1/2} k^{-1/3}, |\dot{c}|)$,

and

(ii) $t_{cf} < M \Delta^{1/2} k^{-1/3}$.

By the energy equation for CF we have

$$\dot{x}^2 \geq \dot{c}^2 + 1 - x^2$$

and so $|\dot{f}| > \max(\Delta^{1/2} k^{-1/3}, |\dot{c}|)$ which gives (i). It also follows that

$$t_{cf} = \int_F^C \frac{dx}{|\dot{x}|} \leq \int_f^1 \frac{dx}{(1 - x^2)^{1/2}} = \text{arc} \cos f \leq \frac{M\Delta^{1/2}}{k^{1/3}}.$$

LEMMA 8. If a solution starts from C and reaches F where $f = 1 - \Delta k^{-2/3}$, $\Delta > 0$, then

(i) $t_{fa'} < M/(\Delta k^{1/3})$,

and

(ii) $|\dot{a}' + \tfrac{2}{3}k - \dot{c}| < t_{cf} + t_{fa'} \leq M\Delta^{1/2}k^{-1/3}$.

Put $x = 1 - \eta$ so that $\ddot{\eta} - 2k(\eta - \tfrac{1}{2}\eta^2)\dot{\eta} - 1 + \eta = 0$. Then $\ddot{\eta} > k\eta\dot{\eta}$ for $0 < \eta < 1$, and integrating from C to A' we have

$$\dot{\eta} > \tfrac{1}{2} k \eta^2 - \dot{c} > \tfrac{1}{2} k(1 - x)^2.$$

Hence

$$t_{fa'} = \int_{A'}^F \frac{dx}{|\dot{x}|} \leq \int_0^f \frac{2\, dx}{k(1-x)^2} \leq \frac{M}{\Delta k^{1/3}}.$$

Part (ii) follows immediately from the integrated equation §2 (3) applied to CA'.

§7. We now jump from A' to the arc AB which is the reflexion of A'B'. Whereas at C the magnitude of $\dot{c}$ made little difference to the subsequent behaviour of the solution, after A the behaviour depends very much on the magnitude of $\dot{a}$. If $\dot{a}$ is very small, the possibility of the solution turning down before it reaches B, or taking a very long time to reach B, cannot be ruled out, but it is of some interest to see that even if $\dot{a} = \underline{O}(k^{-1})$ the subsequent behaviour may be similar to that of the periodic solution for which $\dot{a} \sim \frac{2}{3}k$ except that $h \sim 3^{1/2}$ instead of 2 and that it takes a fairly long time to get away from A.

LEMMA 9. A solution starting from A with $\dot{a} \geq 3/k$ reaches B with

(1) $$|\dot{b} - \dot{a} - \tfrac{2}{3}k| \leq t_{ab} < M\frac{\log k}{k},$$

provided that $k > k_0$. If further $\dot{a} > \varepsilon k$, where $\varepsilon > 0$, then

$$t_{ab} < \frac{1}{2\varepsilon k} + \frac{M}{k} \text{ for } k > k_0.$$

Let G be the point $x = \frac{1}{2}$. Then in AG $\ddot{x} \geq \frac{3}{4}k\dot{x} - x$, and so $\dot{x}$ increases from A as long as $x \leq \frac{3}{4}k\dot{a}$. Since $x \leq \frac{1}{2}$ in AG and $\frac{3}{4}k\dot{a} \geq \frac{9}{4}$, G is reached with $t_{ag} < \frac{1}{2}/\dot{a} \leq \frac{1}{6}k$. On AG by the integrated equation

$$\dot{x} \geq \frac{3}{k} + kx(1 - \tfrac{1}{3}x^2) - \int_A^X x\,dt$$

$$> \frac{3}{k} + \frac{11}{12}kx - \frac{1}{6}kx = \frac{3}{k} + \frac{3}{4}kx.$$

Hence $\dot{g} > \frac{3}{k} + \frac{3}{8}k$, and

$$t_{ag} = \int_A^G \frac{dx}{\dot{x}} < \int_0^{\frac{1}{2}} \frac{dx}{\frac{3}{k} + \frac{3}{4}kx} < \frac{M\log k}{k} \text{ for } k > k_0.$$

Next $\dot{x} > \frac{1}{8}k$ near G; if ever $\dot{x} = \frac{1}{8}k$ on GB, let X be the first occasion. Then

$$t_{gx} < \tfrac{1}{2}/(\tfrac{1}{8}k) = 4/k,$$

but also

$$\dot{x} > \dot{g} - \int_G^X x\,dt > \frac{3}{k} + \frac{3k}{8} - \frac{4}{k} = \frac{3k}{8} - \frac{1}{k} > \frac{1}{8}k$$

for $k > 2$ which gives a contradiction. So $\dot{x} > \frac{1}{8}k$ on GB, B is reached with $t_{gb} < 4/k$ and (1) follows from the integrated equation §2 (3) for AB.

If $\dot{a} > \varepsilon k$, since $\dot{x}$ increases in AG we have $t_{ag} < \frac{1}{2}/\dot{a} < \frac{1}{2}/(\varepsilon k)$, and the further result follows.

This version of Lemma 9 is due to Mr. G. E. H. Reuter. It is shorter and gives a better result for t_{ag} than the original version. He has also shown that the result remains true whenever $\dot{a} \geq \delta/k^2$, $\delta > 0$, $k > k_0(\delta)$. His method is similar, but it involves another subdivision of AB at D where $d = \frac{1}{2}\delta/k$.

§8. From B to H we restrict ourselves to solutions which have a fairly large $\dot{b}$, as there seems no particular interest attaching to those for which $\dot{b}$ is small and only those with large $\dot{b}$ rise to an H for which the hypotheses of Lemmas 3, 4 and 5 are satisfied.

LEMMA 10. If $0 < \delta < 1$ and $0 < \varepsilon < \frac{2}{3}(1 - \delta)$ a solution starting from B with $\frac{2}{3}k(1 - \delta) < \dot{b} < M_0 k$ reaches a point P in BH at which $\dot{p} = \varepsilon k$ with $t_{bp} < M/(\varepsilon k)$, $p \geq \frac{3}{2}$, and

(1) $$|\dot{b} + k(p - p^3/3 - 2/3)| \leq 2\ \varepsilon k\ ,$$

for $\delta < \delta_0$, $\varepsilon < \varepsilon_0$ and $k > k_0(\varepsilon, \delta)$.

Since $\ddot{x}$ is obviously negative in BH, $\dot{x}$ decreases and so P is uniquely defined, and $t_{bp} < (p - 1)/(\varepsilon k)$. From the integrated equation we have

(2) $$\int_B^P \ddot{x}\ dt = \dot{b} - \varepsilon k + k(p - p^3/3 - 2/3)\ ,$$

and since $\dot{b} < M_0 k$, the right hand side is negative for $p > M$ while the left hand side is positive. Hence $p < M$ and so $t_{bp} < M/(\varepsilon k)$. Using (2) again, we have

$$|\dot{b} + k(p - p^3/3 - 2/3)| = \varepsilon k + \int_B^P \frac{\ddot{x}\ dx}{\dot{x}} \leq \varepsilon k + \frac{M}{\varepsilon k} \leq 2\ \varepsilon k$$

for $k > k_0(\varepsilon)$. Since $\dot{b} > \frac{2}{3} k(1 - \delta)$, this gives

$$p - \frac{p^3}{3} - \frac{2}{3} < - \frac{2}{3}(1 - \delta) + 2\varepsilon$$

i.e.

$$p - p^3/3 < 2\ \delta + 2\varepsilon$$

and so $p^2 > 3 - M\delta - M\varepsilon > 2\frac{1}{4}$ for $\delta < \delta_0$, $\varepsilon < \varepsilon_0$ which completes the proof.

LEMMA 11. If the conditions of Lemma 10 are satisfied, there is a point Q in PH such that $\dot{q} = 1/(\varepsilon k)$ and $t_{pq} < (3 \log k)/k$ for $\delta < \delta_0$ $\varepsilon < \varepsilon_0$, $k > k_0(\varepsilon, \delta)$.

Since $\ddot{x} < 0$ in PH the existence of Q is obvious for $k > \varepsilon^{-1}$. Since $x^2 - 1 \geq p^2 - 1 > 1$ in PQ, we have

$$\ddot{x} = - k\ \dot{x}(x^2 - 1) - x \leq - k\ \dot{x}\ ,$$

and so dividing by $\dot{x}$ and integrating from P to Q we obtain

$$\log(\dot{q}/\dot{p}) \leq - k\ t_{pq}\ .$$

Hence

$$k\ t_{pq} \leq 2\ \log(\varepsilon k)$$

from which the result for t_{pq} follows. The restrictions on δ, ε and k are mainly needed to ensure that $p^2 - 1 > 1$.

LEMMA 12. If the hypotheses of Lemmas 10 and 11 are satisfied, then $t_{qh} < M/(\varepsilon k)$ and

$$|\dot{b} + k(h - \tfrac{1}{3} h^3 - \tfrac{2}{3})| \leq M\frac{\log k}{k}$$

for $\varepsilon < \varepsilon_0$, $\delta < \delta_0$ and $k > k_0(\varepsilon, \delta)$.

$\ddot{x} < - x$ and as above $x < M$. Since $h = 0$

$$- \frac{1}{\varepsilon k} = \int_Q^H \ddot{x}\ dt < - M\ t_{qh}$$

from which the result for t_{qh} follows.

By the integrated equation for BH we have

$$|\dot{b} - k\,(h - \tfrac{1}{3}h^3 - \tfrac{2}{3})| = \int_B^H x\,dt \lesssim M\,t_{bh} \lesssim M\frac{\log k}{k}$$

for $\delta < \delta_0$, $\varepsilon < \varepsilon_0$ and $k > k_0(\varepsilon, \delta)$.

§9. We have now covered the equivalent of a half wave HCA'B'H', and we may review the results as follows: in Lemma 1 we showed that a solution starting at a maximum H not too far above $x = 1$ arrives at C with $|\dot{c}|$ bounded by a constant depending on h. In Lemma 2 we showed that although $\ddot{x} \lesssim 0$ in HC, $\dot{x}$ remains small until x approaches 1, and consequently the solution takes a long time to reach C unless h is near 1. Lemmas 3, 4, and 5 show that the one-sided estimates for $\dot{x}$ and for the time in Lemma 2 do in fact give a good approximation, provided that $h > \frac{3}{2}$ and x is not too near 1. Lemma 6 deals with the arc just before C, and Lemma 7 with that just after C assuming that the solution starts from C, and together they show that the time taken to cross the strip $|x - 1| \lesssim \Delta k^{-2/3}$ is at most $M\Delta^2 k^{-1/3}$. Lemma 8 shows that a solution starting from C reaches A' in time $M\Delta^{1/2} k^{-1/3}$ with $|\dot{a}'| > \frac{2}{3}k + |\dot{c}| - M\ k^{-\frac{1}{3}}$. Lemma 9 shows that a solution starting from A with $\dot{a} > 3/k$ rises to B in time $O(\log k/k)$ with $\dot{b} > \frac{2}{3}k - \underline{o}\,(k)$ and gives a better result if $\dot{a}$ is large. Lemmas 10, 11, and 12 deal with a solution rising from B with $\dot{b}$ large. They show that $t_{bh} = \underline{O}\,(\log k/k)$, and give a formula for h depending on $\dot{b}$. It may be observed that the transition interval PQ is comparable in some ways to the transition interval YZ in particular the time for each is $O(\log k/k)$.

§10. The periodic solution traces the reflection of the half wave ABHCA' below $x = 0$, and we shall now proceed to estimate T and h by applying the previous lemmas with this in mind.

THEOREM 1. For the periodic solution

(i) $\dfrac{M}{\Delta k^{1/3}} \lesssim |\dot{c}| \lesssim \dfrac{M\Delta^{1/2}}{k^{1/3}}$,

(ii) $|\dot{a} - \frac{2}{3}k| \lesssim M\Delta^{1/2}k^{-1/3}$

(iii) $|\dot{b} - \frac{4}{3}k| < M\Delta^{1/2}k^{-1/3}$,

(iv) $|h - 2| < M\Delta^{1/2}k^{-4/3}$,

(v) $|T - k\{\frac{3}{2} - \log_e 2\}| < M\Delta^2 k^{-1/3}$,

provided that in each case $\Delta > \Delta_0$, $k > k_0(\Delta)$.

Part (i) follows immediately from Lemma 6, and putting this in Lemma 8 and reflecting in $x = 0$, we have (ii). Then since $\dot{a} > \frac{1}{3}k$ for $k > k_0$, we can use the last part of Lemma 9 which with (ii) gives (iii). Using this in Lemma 12, we have

$$|h - \tfrac{1}{3}h^3 + \tfrac{2}{3}| \lesssim M\Delta^{1/2}k^{-4/3},$$

and so if we write $h = 2 + \zeta$ we have

$$|\zeta\,(3 + 2\zeta + \tfrac{1}{3}\zeta^2)| \lesssim M\Delta^{1/2}k^{-4/3},$$

which gives (iv) provided that $\Delta > \Delta_0$ and $k > k_0(\Delta)$.

Finally

$$T = t_{ah} + t_{hc} + t_{ca'}$$

By Lemmas 9, 10, 11 and 12

$$t_{ah} = t_{ag} + t_{gb} + t_{bp} + t_{pq} + t_{qh}$$
$$\leq \frac{M}{k} + \frac{M}{\varepsilon k} + \frac{M \log k}{k} + \frac{M}{\varepsilon k} \leq \frac{M \log k}{k}$$

for every $\varepsilon > 0$, provided that $k > k_0(\varepsilon)$. By Lemmas 7 and 8 $t_{ca'} \leq M \Delta^{1/2} k^{-1/3}$, and using (iv) in Lemma 2 we have

$$t_{hc} \geq k \left\{ \frac{3}{2} - \log 2 \right\} - M \Delta^{1/2} k^{-1/3},$$

and similarly from Lemmas 3, 4, 5 and 6, putting $e = 1 + \Delta k^{-2/3}$ we have, as in §5,

$$t_{hc} = t_{hy} + t_{yz} + t_{ze} + t_{ec}$$
$$\leq \frac{M}{\delta k} + \frac{M \log k}{k} + k \left\{ \frac{3}{2} - \log 2 \right\} - \frac{\Delta^2}{k^{1/3}} + \frac{M \Delta^2}{k^{1/3}}$$
$$\leq k \left\{ \frac{3}{2} - \log 2 \right\} + \frac{M \Delta^2}{k^{1/3}}$$

provided that $\Delta > \Delta_0$, $k > k_0(\Delta, \delta) = k_0(\Delta)$.

§11. It remains to remove the large constant Δ and replace it by constants which can be determined more precisely. It is obvious that the errors depending on Δ originate in the arc EF, i.e. in the strip $|x - 1| \leq \Delta k^{-2/3}$ with $\dot{x} < 0$, and by Lemmas 6 and 7

(1) $$\frac{M}{\Delta k^{1/3}} \leq |\dot{x}| \leq \frac{M \Delta^{1/2}}{k^{1/3}}$$

on this arc. We therefore put

$$1 - x = k^{-2/3} \xi, \quad - \dot{x} = k^{-1/3} \eta,$$

so that

$$- \ddot{x} = \eta \frac{d\eta}{d\xi},$$

and

(2) $$\eta \frac{d\eta}{d\xi} - 2 \xi \eta - 1 + k^{-2/3}(\xi^2 \eta + \dot{\xi}) = 0,$$

where $0 < \eta < M \Delta^{1/2}$ and $- \Delta < \xi < \Delta$ in EF.

It is evident from (1) and (2) that the behaviour of solutions near $x = 1$ will be similar to that which is given by the simpler equation

(3) $$\eta_0 \frac{d\eta_0}{d\xi} - 2 \xi \eta_0 - 1 = 0$$

with

(4) $$\frac{M}{\Delta} \leq \eta_0(0) \leq M \Delta^{1/2}$$

We therefore proceed to investigate (3).

LEMMA 13. Solutions of (3) satisfying (4) may be divided into two classes: one for which $\eta_0(\xi)/\xi^2 \to 1$ as $\xi \to -\infty$ and one for which $\eta_0(\xi) \to 0$ as ξ tends to some finite negative value ξ_0. These two sets of solutions are separated by a unique solution for which $\eta_0^*(0) = \alpha$, and $2\xi\eta_0^* \to -1$ as $\xi \to -\infty$, and $\eta_0^*(\xi)/\xi^2 \to 1$ as $\xi \to +\infty$.

By drawing the hyperbola $2\xi\eta_0 + 1 = 0$, we see that $d\eta_0/d\xi > 0$ for $\eta_0 > 0$ below and to the right of the upper branch and $d\eta_0/d\xi < 0$ above this branch. Also $d\eta_0/d\xi \to \infty$ as $\xi \to \pm\infty$ unless $2\xi\eta_0 + 1 \to 0$. Hence one set of solutions descends from $+\infty$ as ξ increases from $-\infty$ until it crosses the upper branch of the hyperbola and then ascends again to $+\infty$, another set

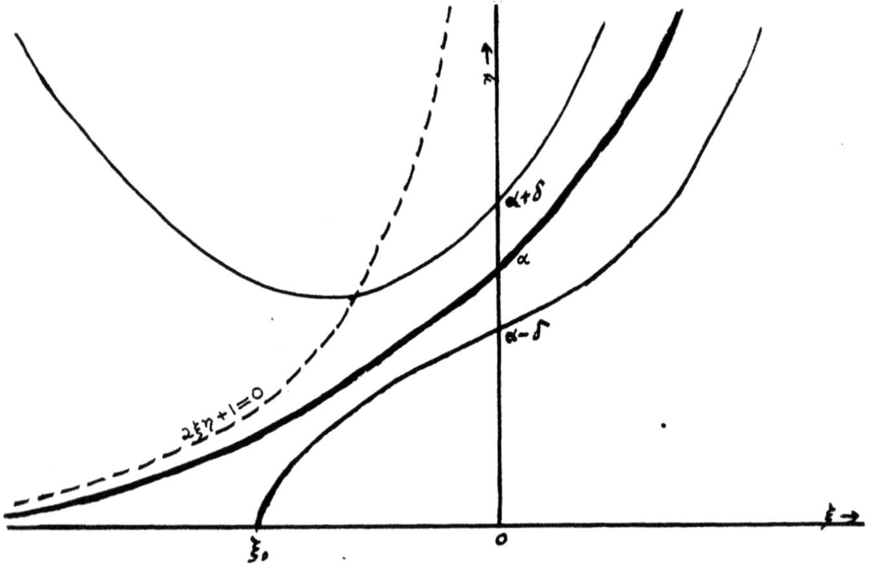

Figure 2

The curves $\eta_0(\xi)$ are shown in black with the separating curve η_0^* on which $\eta_0(\xi) \to 0$ as $\xi \to -\infty$ thicker than the others.

crosses $\eta_0 = 0$ at $\xi = \xi_0$, say, with $d\eta_0/d\xi = \infty$ at $\xi = \xi_0$ and ascends to $+\infty$. If $\xi \to \pm\infty$ in such a way that $\underline{\lim}\ \eta_0 > 0$, then

$$\frac{d\eta_0}{d\xi} = 2\xi + \frac{1}{\eta} \sim 2\xi$$

and so $\eta_0(\xi) \sim \xi^2$ as $\xi \to \pm\infty$. Both sets vary continuously with value of $\eta_0(0)$ and form open sets. There is therefore a solution, or closed set of solutions, separating the two sets, and it must lie between the hyperbola and $\eta_0 = 0$ for negative ξ. It follows that on such a solution $\eta_0^*\, d\eta_0^*/d\xi > 0$, and since for fixed ξ $d\eta_0/d\xi$ increases as η_0 decreases, the lower solutions increase more rapidly than the upper which is impossible if there is more than one solution on which $\xi \to -\infty$ between the hyperbola and $\eta_0 = 0$. Hence there is a unique solution separating the classes.

LEMMA 14. Suppose that $\eta(\xi)$ satisfies (2) and that α is defined as in Lemma 13 (or §1 (5)). Then if $\eta(0) \geq \alpha + \delta$ where $\delta > 0$, we have $\eta \geq 1$ for some ξ such that $-\Delta \leq \xi < 0$; if $\eta(0) \leq \alpha - \delta$, where $0 < \delta < \alpha$, then $\eta = 0$ for some ξ such that $-\Delta \leq \xi < 0$ provided that in both cases $\Delta > \Delta_0(\delta)$, and $k > k_0(\delta, \Delta)$.

Let η be a solution of (2) such that $\eta(0) = \alpha + \zeta \geq \alpha + \delta$ and let $\eta = \eta_0 + \eta_1$, where η_0 is a solution of (3) such that $\eta_0(0) = \alpha + \zeta$. Then by Lemma 13 $\eta_0(\xi) > 0$ for all ξ and $\eta_0(\xi) \to \infty$ as $\xi \to -\infty$, and so $\eta_0(\xi) \geq 2$ for $\xi \leq -\Delta$, provided that $\Delta > \Delta_0(\delta)$, and also $\min \eta_0 \geq \varepsilon(\delta) > 0$. Substituting in (2) and using (3), we have

$$(5) \qquad (\eta_0 + \eta_1)\frac{d\eta_1}{d\xi} + \eta_1 \frac{d\eta_0}{d\xi} - 2\xi\eta_1 + \frac{1}{k^{2/3}}\{\xi^2(\eta_0 + \eta_1) + \xi\} = 0,$$

where $\eta_1(0) = 0$, and since $\eta_0(\xi) \geq \varepsilon > 0$, we have

$$\frac{d\eta_1}{d\xi} \leq \frac{\chi(\Delta)|\eta_1|}{\varepsilon - |\eta_1|} + \frac{1}{k^{2/3}}\left\{\xi^2 + \frac{|\xi|}{\varepsilon - |\eta_1|}\right\}$$

where $\chi(\Delta)$ denotes a number depending on Δ but not on k or ξ. Since $\eta_1(0) = 0$ by a well known method[6] it follows that for every $\Delta > 0$

$$(6) \qquad |\eta_1| \leq \chi(\Delta)\, k^{-2/3} < \ /4,$$

for $-\Delta \leq \xi \leq \Delta$, provided that $|\eta_1| \leq \ /2$, and $k > k_0(\varepsilon, \Delta)$. Hence (6) holds for every $\delta > 0$ and every $\Delta > 0$, provided that $-\Delta \leq \xi \leq \Delta$ and $k > k_0(\delta, \Delta)$. But this gives $|\eta_1| \leq 1$ and so $\eta \geq \eta_0 - |\eta_1| \geq 1$ for $\xi = -\Delta$ provided that $\Delta > \Delta_0$ and $k > k_0(\delta, \Delta)$.

If $\eta_0(0) = \alpha - \delta/2$ where $0 < \delta < \alpha$, we have $\eta_0(\xi) = 0$ for some $\xi = \xi_0 > -\Delta$ provided that $\Delta > \Delta_0(\delta)$. Now if $\xi < 0$, $\eta > 0$ and $2\xi\eta + 1 > 0$ we also have $\xi(\xi\eta + 1) < 0$, and so if η satisfies (2)

$$\eta\frac{d\eta}{d\xi} > 2\xi\eta + 1.$$

This means that the solution of (2) increases more rapidly than the solution of (3) through the same point ξ, η in the region defined by $\xi < 0$, $\eta > 0$, $2\xi\eta + 1 > 0$. The solution of (2) through $\xi = 0$, $\eta = \alpha - \delta$ certainly lies below the $\eta_0(\xi)$ just defined near $\xi = 0$, and continues to lie below it as ξ decreases. It therefore reaches $\eta = 0$ for $\xi > \xi_0$ which gives the second part.

LEMMA 15. Suppose that α and β are defined by §1 (5) and ε is any positive number. Then if $\eta(\xi)$ is a solution of (2) for which $|\eta(0) - \alpha| < \delta$,

$$\left|\int_{-\Delta}^{0}\frac{d\xi}{\eta} - \Delta^2 - \alpha\right| < \varepsilon, \quad \left|\int_{0}^{\Delta}\frac{d\xi}{\eta} - \beta\right| < \varepsilon,$$

(6) See E. Kamke, Differentialgleichungen reeller Funktionen, (Leipzig 1930, New York 1947), 93.

provided that $\Delta > \Delta_0(\varepsilon)$, $\delta < \delta_0(\varepsilon,\Delta)$ and $k > k_0(\varepsilon,\delta,\Delta)$.

We first observe that since $\eta_0^* \sim \xi^2$ as $\xi \to +\infty$ the integral in §1 (5) is convergent so that β is defined. Suppose that now η_0^* is defined by §1 (4) and §1 (5) and let $\eta = \eta_0^* + \eta_1$, then $\min \eta_0^* \geq \zeta(\Delta) > 0$ for $-\Delta \leq \xi \leq \Delta$, and so, as in (6), for every $\varepsilon' > 0$ and every $\Delta > 0$ we have $|\eta_1| < \varepsilon'$ for $-\Delta \leq \xi \leq \Delta$ provided that $\delta < \delta_0(\varepsilon',\Delta)$, $k > k_0(\delta,\varepsilon',\Delta)$. Hence

$$\left| \int_0^\Delta \frac{d\xi}{\eta} - \beta \right| \leq \int_0^\Delta \frac{|\eta_1| d\xi}{\eta \cdot \eta_0^*} + \int_\Delta^\infty \frac{d\xi}{\eta_0^*}$$

$$\leq \frac{\varepsilon'}{\alpha - \delta} \int_0^\Delta \frac{d\xi}{\eta_0^*} + \int_\Delta^\infty \frac{d\xi}{\eta_0^*} < \varepsilon$$

for every $\varepsilon > 0$, provided that $\Delta > \Delta_0(\varepsilon)$, $\varepsilon' < \varepsilon_0'(\varepsilon,\Delta)$, $\delta < \delta_0(\varepsilon',\Delta)$, and $k > k_0(\varepsilon, \varepsilon', \Delta, \delta)$.

As $\xi \to -\infty$, η_0^* tends to 0, and by Lemma 2

$$\eta(-\Delta) = -\dot{e}\, k^{1/3} \leq \frac{h}{2\Delta},$$

and so, dividing (2) by η and integrating and writing

$$I(\Delta) = \int_{-\Delta}^0 \frac{d\xi}{\eta}$$

we have

$$-\eta(-\Delta) + \eta(0) = -\Delta^2 + I(\Delta) - \frac{1}{k^{2/3}}\left\{\frac{\Delta^3}{3} + \int_{-\Delta}^0 \frac{\xi d\xi}{\eta}\right\}.$$

It follows that

$$|I(\Delta) - \Delta^2 - \alpha| \leq \frac{h}{2\Delta} + |\alpha - \eta(0)| + \frac{\Delta^3}{3k^{2/3}} + \frac{\Delta I(\Delta)}{k^{2/3}}.$$

By choosing $\delta < \delta_0$, $\Delta > \Delta_0$ and then $k > k_0(\varepsilon,\delta,\Delta)$ we can make

$$I(\Delta) < \Delta^2 + \alpha + (1/2)\, I(\Delta) + 1$$

and so $I(\Delta) \leq 2(\Delta^2 + \alpha + 1)$. Then for every $\varepsilon > 0$ and every $\Delta > \Delta_0$ we have

$$|I(\Delta) - \Delta^2 - \alpha| < \varepsilon$$

for $\delta < \delta_0$, $\Delta > \Delta_0$ and $k > k_0(\varepsilon,\delta,\Delta)$ which completes the result.

§12. We now apply Lemmas 14 and 15 to the periodic solution.

THEOREM 2. For the periodic solution as $k \to \infty$

(i) $\dot{c} = -\alpha k^{-1/3} + \underline{o}(k^{-1/3})$,

(ii) $\dot{a} = \frac{2}{3} k + (\alpha + \beta) k^{-1/3} + \underline{o}(k^{-1/3})$,

(iii) $\dot{b} = \frac{4}{3} k + (\alpha + \beta) k^{-1/3} + \underline{o}(k^{-1/3})$,

(iv) $h = 2 + \frac{1}{3}(\alpha + \beta) k^{-4/3} + \underline{o}(k^{-4/3})$,

(v) $T = k\{\frac{3}{2} - \log 2\} + \frac{3}{2}(\alpha + \beta) k^{-1/3} + \underline{o}(k^{-1/3})$

where α and β are defined by §1 (5).

By Lemma 2 $|x| \leq M k^{-1/3}/\Delta$ for $x \geq 1 + \Delta k^{-2/3}$, and so in the notation of Lemma 14 $0 < \eta < M/\Delta$ for $\xi = -\Delta$. If $|\dot{c}\, k^{1/3} + \alpha| \geq \delta$ the result of Lemma 14 gives a contradiction for $\Delta > \Delta_0(\delta)$ and $k > k_0(\delta, \Delta)$, and so $|\dot{c}\, k^{1/3} + \alpha| \leq \delta$ for every $\delta > 0$ and $k > k_0(\delta)$.

By the integrated equation for CA'

$$|\dot{a}' - \dot{c} + \tfrac{2}{3} k + t_{cf}| \leq \int_F^{A'} x\, dt + \int_C^F (1 - x) dt.$$

Now

(1)
$$t_{cf} = \int_F^C \frac{dx}{\dot{x}} = \frac{1}{k^{1/3}} \int_0^\Delta \frac{d\xi}{\eta},$$

and so by Lemmas 7, 8 and 15 we have

$$|\dot{a}' - \dot{c} + \tfrac{2}{3} k + \beta k^{-1/3}| < \frac{\varepsilon}{k^{1/3}} + \frac{M}{\Delta k^{1/3}} + \frac{\Delta}{k^{2/3}} \cdot \frac{M \Delta^{1/2}}{k^{1/3}} < \frac{2\varepsilon}{k^{1/3}}$$

for every $\varepsilon > 0$, provided that $\Delta > \Delta_0(\varepsilon)$ and $k > k_0(\varepsilon, \Delta)$. Reflecting in $x = 0$ we have (ii). Then (iii) follows from Lemma 9. Putting (iii) in Lemma 12 we have

$$h - \tfrac{1}{3} h^3 + \tfrac{2}{3} + (\alpha + \beta)\, k^{-4/3} = \underline{o}(k^{-4/3}),$$

and putting $h = 2 + \varphi$, we have (iv) as in Theorem 1.

As in Theorem 1 $t_{ah} = O(\log k/k)$. By Lemma 2 (iv) and (iv) above we have

(2) $$t_{he} > k\,(\tfrac{3}{2} - \log 2) + \tfrac{1}{2}(\alpha + \beta)\, k^{-1/3} - k(e - 1)^2 - k^{-1/3} - M\Delta^3 k^{-1}$$

for every $\varepsilon > 0$, $\Delta > 0$ and $k > k_0(\varepsilon, \Delta)$. By Lemmas 3, 4, 5 with (iv) above and (1)

(3)
$$\begin{aligned} t_{he} &= t_{hy} + t_{yz} + t_{ze} \\ &\leq \frac{M}{\delta k} + \frac{M \log k}{k} + k(\tfrac{3}{2} - \log 2) + \tfrac{1}{2} \frac{(\alpha + \beta)}{k^{1/3}} \\ &\quad + \frac{\varepsilon}{k^{1/3}} - k(e - 1)^2 \end{aligned}$$

for every $\varepsilon > 0$, provided that $k > k_0(\varepsilon)$. But

$$t_{ec} = \int_C^E \frac{dx}{|\dot{x}|} = \frac{1}{k^{1/3}} \int_{-\Delta}^0 \frac{d\xi}{\eta},$$

and so by Lemma 15 and (i) above we have

(4) $$|t_{ec} - k(e - 1)^2 - \alpha k^{-1/3}| < \varepsilon k^{-1/3}$$

for every $\varepsilon > 0$, provided that $\Delta > \Delta_0(\varepsilon)$ and $k > k_0(\varepsilon, \delta, \Delta)$.

Putting (2) and (3) together with (4), we have

$$t_{hc} = k(\tfrac{3}{2} - \log 2) + \tfrac{1}{2}(\alpha + \beta)\, k^{-1/3} + \alpha k^{-1/3} + o(k^{-1/3})$$

as $k \to \infty$. Combining this with (1) using Lemma 15 again, and the lemmas which cover ABH and CA' we have (v).

§13. Mr. Reuter has also drawn my attention to the fact that

$$\eta_0 \frac{d\eta_0}{d\xi} = 2\xi\eta_0 + 1$$

can be solved explicitly by means of tables. Put $\eta_0 = \xi^2 + u$, then

$$(\xi^2 + u)\frac{du}{d\xi} = 1, \text{ and } \frac{d\xi}{du} = \xi^2 + u\ .$$

The latter is a Riccati equation; put $\xi = -\frac{1}{\varphi}\frac{d\varphi}{du}$, then

$$\frac{d^2\varphi}{du^2} + u\varphi = 0.$$

To get the solution η_0^* for which $\eta_0^* \to 0$ as $\xi \to -\infty$ we must have $u \sim -\xi^2$, $\xi \sim -|u|^{1/2}$. Hence $\log \varphi \sim -(2/3)|u|^{3/2}$ as $u \to -\infty$. The correct solution is

$$\varphi = A_i(-u)$$

in the standard notation for Airy integrals (B. A. Math. Tables Part Volume B 1946). Now $Ai(-u)$ is exponentially small as $u \to -\infty$, rises to a maximum at $u = u_1 = 1.019 \ldots$, and becomes zero at $u = u_2 = 2.338 \ldots$. At $u = u_1$, $\varphi' = 0$ so that $\xi = 0$ and therefore $\alpha = u_1$. Also

$$\beta = \int_0^\infty \frac{d\xi}{\eta_0^*} = \int_0^\infty \frac{d\xi}{\xi^2 + u} = \int_{u_1}^{u_2} du = u_2 - u_1$$

since $\xi \to +\infty$ as $u \to u_2$. Hence $\alpha + \beta = u_2 = 2.338$ and

$$2T = 2k(\tfrac{3}{2} - \log 2) + 7.014k^{-1/3} + o(k^{-1/3})$$

as in Dorodnitsin's paper* (although of course he gives further terms).

* See also J. J. Stoker, Nonlinear Vibrations (New York 1950) 141.

PERTURBATIONS OF LINEAR SYSTEMS WITH CONSTANT COEFFICIENTS POSSESSING PERIODIC SOLUTIONS*

By E. A. Coddington and N. Levinson

INTRODUCTION

Let x, f be real vectors with n components x_i, f_i $(i = 1,\ldots,n)$ respectively, and suppose t and μ are real. Consider the system of differential equations

$$(0.1) \qquad x' = f(x, t, \mu), \quad (' = d/dt) ,$$

where f, $\partial f/\partial x_i$ are continuous in (x, t, μ) for small $|\mu|$ and (x, t) in some region V. The situation of interest here is when the system $x' = f(x, t, 0)$ has a periodic solution $x = p(t)$ of period T, and f is of period T in t. If g is any differentiable n-dimensional vector function of x, denote by g_x the matrix with element $\partial g_i/\partial x_j$ in the i-th row and j-th column. Then the system of the first variation of (0.1) with respect to this solution p is

$$(0.2) \qquad \xi' = f_x(p(t), t, 0)\,\xi .$$

As is well known, if (0.2) has no solution of period T, then (0.1) has a periodic solution of period T for sufficiently small $|\mu|$, $x = q(t,\mu)$, which is continuous in (t, μ), and such that $q(t, 0) = p(t)$.

The fact that (0.2) has no periodic solution of period T is equivalent to the non-vanishing of a certain Jacobian. Let $x = x(t, \mu, \eta)$ be a solution of (0.1) with $x(0, \mu, \eta) = p(0) + \eta$. Then the condition for periodicity is just

$$(0.3) \qquad x(T, \mu, \eta) - p(0) - \eta = 0 .$$

For $\mu = 0$, the system (0.3) has $\eta = 0$ as a solution. If the Jacobian $J = \det\,(x_\eta(T,\mu,\eta) - E)$ does not vanish at $\mu = 0$, then (0.3) has a unique continuous solution $\eta = \eta(\mu)$ for $|\mu|$ sufficiently small with $\eta(0) = 0$. Here det = determinant, and E represents the n-dimensional unit matrix.

In case $f(x, t, \mu)$ does not contain t explicitly, but is of the form $f(x,\mu)$, the system (0.2) has $\xi = p'$ as a solution of period T so that the above hypothesis is never fulfilled. In this case the theorem is modified, and if

$$(0.4) \qquad x' = f(x, \mu)$$

* This paper was written in the course of work sponsored by the Office of Naval Research.

has a periodic solution $x = p(t)$ of period T_0 for $\mu = 0$, and if the first variation (0.2) has no more than one independent solution of period T_0, then (0.4) has a periodic solution $x = q(t,\mu)$ for small $|\mu|$, continuous in (t, μ). The period, $T(\mu)$, of $q(t,\mu)$ is continuous in μ. Moreover $q(t, 0) = p(t)$, and $T(0) = T_0$. The fact that (0.4) at $\mu = 0$ has p' as its only linearly independent solution of period T_0 is equivalent to the non-vanishing of a certain $(n - 1)$-dimensional Jacobian J_1.

In many important cases the hypothesis on the equations of the first variation, or, what is the same, the non-vanishing requirement on the Jacobians J and J_1, is not met. For example in case w is a scalar the equation

$$w'' + w = \mu g(w, w', t, \mu) , \tag{0.5}$$

where g is periodic of period 2π in t, if it contains t, has for its first variation

$$w'' + w = 0 ,$$

which has sin t and cos t as independent solutions of period 2π. Thus the hypothesis for neither result stated above can be met for (0.5). The cases of equation (0.5) relevant here have been treated in detail by Friedrichs and Stoker [2].

It is our purpose to consider the case where $f(x, t, 0)$ of (0.1), or $f(x, 0)$ of (0.4), are of the form Ax, where A is a constant matrix. We show that it is possible to give sufficient conditions for the existence of periodic solutions of $x' = A x + \mu f(x, t,\mu)$ (and $x' = A x + \mu f(x, \mu)$) for small $|\mu|$ even when the Jacobians J and J_1 vanish. These conditions only involve knowledge of the solutions of the degenerate linear system $x' = A x$. As is to be expected, if f is analytic, then the solutions (and period in case $f = f(x, \mu)$) can be solved for recursively. In the following, systems for which J (or J_1) vanish will be referred to simply as systems with a vanishing Jacobian.

1. PERTURBATION OF A SYSTEM WITH A VANISHING JACOBIAN

For the linear system

$$x' = Ax, \tag{1.1}$$

where A is a constant real matrix, assume that there exists a real periodic solution $p = p(t)$ with period 2π. This is equivalent to the fact that there exists at least one characteristic root λ of A of the form $\lambda = iN$ where N is an integer (which may be zero). We shall be interested in the perturbed system

$$x' = Ax + \mu f(x, t, \mu) , \tag{1.2}$$

where it is assumed that A is the constant matrix given in (1.1), μ is a real parameter, and f is real, periodic of period 2π in t. (Since 2π need not be the least period of f in t the case of subharmonic oscillations is not excluded).

It is clear that if c and d are any real constants, then $c\,p(t+d)$ is also a real periodic solution of (1.1) with period 2π. It is not obvious for what values of c and d, if any, the perturbed system (1.2) may have a periodic solution tending to $c\,p(t+d)$ as $\mu \to 0$. We can not apply the procedure mentioned in the Introduction for in the case (1.2) the relevant Jacobian J vanishes. However, it is possible to give sufficient conditions for the existence of periodic solutions of (1.2) for $\mu \neq 0$ provided A is in canonical form, and it is always possible to arrange this.

Setting $x = Py$ where P is a real non-singular constant matrix, the system (1.2) can be replaced by a system for y where the coefficient matrix $B = P^{-1}A\,P$, when $\mu = 0$, is in real canonical form. Moreover, this new system satisfies the same assumptions as (1.2). It will therefore be assumed that A already has the following real canonical form

$$(1.3) \qquad A = \begin{pmatrix} A_1 & & & & & & & \\ & A_2 & & & & & & \\ & & \ddots & & & & & \\ & & & A_k & & & & \\ & & & & B_1 & & & \\ & & & & & B_2 & & \\ & & & & & & \ddots & \\ & & & & & & & B_{m_C} \end{pmatrix},$$

where the elements not shown are zeros. Each A_j, $j = 1, \ldots, k$, is a matrix of α_j (α_j even) rows and columns of the form

$$(1.4) \qquad A_j = \begin{pmatrix} S_j & & & \\ E_2 & S_j & & \\ & \ddots & \ddots & \\ & & E_2 & S_j \end{pmatrix},$$

where all elements are zero except S_j and E_2, and

$$S_j = \begin{pmatrix} 0 & -N_j \\ N_j & 0 \end{pmatrix}, \quad E_2 = \begin{pmatrix} 1 & 0 \\ 0 & 1 \end{pmatrix},$$

N_j being a positive integer. (In the following E_k will always denote the k-dimensional unit matrix $0 < k < n$, and $E = E_n$). A matrix A_j may have only two rows and columns in which case it is S_j. Each matrix B_j has β_j rows and columns, $j = 1, \ldots, m$, and is of the form

$$(1.5) \qquad B_j = \begin{pmatrix} 0 & 0 & \cdots & \cdots & 0 \\ 1 & 0 & & & \vdots \\ 0 & 1 & \ddots & & \vdots \\ \vdots & \ddots & \ddots & \ddots & \vdots \\ 0 & \cdots & 0 & 1 & 0 \end{pmatrix},$$

where B_j may have only one row and column, in which case B_j consists of the single element 0. The matrix C has $\gamma = n - \sum_{j=1}^{k} \alpha_j - \sum_{j=1}^{m} \beta_j$ rows and columns, and has no characteristic roots of the form iN for any integer N, including N = 0. It is useful to notice that C need not be in canonical form.

A fundamental matrix for (1.1) is given by

(1.6) $$e^{tA} = \begin{pmatrix} e^{tA_1} & & & & & & & \\ & e^{tA_2} & & & & & & \\ & & \ddots & & & & & \\ & & & e^{tA_k} & & & & \\ & & & & e^{tB_1} & & & \\ & & & & & e^{tB_2} & & \\ & & & & & & \ddots & \\ & & & & & & & e^{tB_m} \\ & & & & & & & & e^{tC} \end{pmatrix} .$$

Here

(1.7) $$e^{tA_j} = \begin{pmatrix} e^{tS_j} & 0_2 & \cdots & \cdots & 0_2 \\ te^{tS_j} & e^{tS_j} & \ddots & & \vdots \\ \vdots & \ddots & \ddots & \ddots & \vdots \\ \vdots & & \ddots & \ddots & 0_2 \\ \frac{t^{p_j-1} e^{tS_j}}{(p_j-1)!} & \cdots & \cdots & te^{tS_j} & e^{tS_j} \end{pmatrix} ,$$

where $\alpha_j = 2p_j$, 0_2 is the 2 by 2 zero matrix, and

(1.8) $$e^{tS_j} = \begin{pmatrix} \cos N_j t & -\sin N_j t \\ \sin N_j t & \cos N_j t \end{pmatrix} ,$$

while

(1.9) $$e^{tB_j} = \begin{pmatrix} 1 & 0 & \cdots & \cdots & \cdots & 0 \\ t & 1 & 0 & & & \vdots \\ \frac{t^2}{2!} & t & \ddots & \ddots & & \vdots \\ \vdots & & \ddots & \ddots & \ddots & \vdots \\ \vdots & & & \ddots & \ddots & 0 \\ \frac{t^{\beta_j-1}}{(\beta_j - 1)!} & \cdots & \cdots & \cdots & t & 1 \end{pmatrix} .$$

Concerning the matrix e^{tC}, the fact that the characteristic roots of C are not of the form iN, for N an integer, implies that

(1.10) $$\det(e^{2\pi C} - E_\gamma) \neq 0.$$

1.1 <u>Existence of periodic solutions for small $|\mu|$</u>.

Suppose now that (1.2) has a unique solution $x = x(t, \mu, c)$, where $x(0, \mu, c) = c$, which exists for t in some interval containing $0 \leqq t \leqq 2\pi$, and is continuous in μ for μ sufficiently near $\mu = 0$. From (1.2), using the variation of constants formula,

$$x(t,\mu, c) = e^{tA} c + \mu \int_0^t e^{(t-s)A} f(x(s,\mu, c), s, \mu)ds. \tag{1.11}$$

It follows directly from (1.11), and uniqueness, that a necessary and sufficient condition for x to be periodic of period 2π is that

$$(e^{2\pi A} - E) c + \mu \int_0^{2\pi} e^{(2\pi-s)A} f(x(x,\mu, c), s,\mu)ds = 0. \tag{1.12}$$

This represents a system of n equations for the n unknowns consisting of the components of c. In order to state sufficient conditions for the existence of c we first analyze the structure of (1.12) in more detail.

If x is periodic of period 2π and if $c = c_\mu$ is known to exist as a function of μ and be continuous for small $|\mu|$ (or indeed if it exists in the open interval $0 < \mu < \delta$ for some δ and the lim c_μ as $\mu \to 0+$ exists) then from (1.12), letting $\mu \to 0$,

$$(e^{2\pi A}-E)c_0 = 0. \tag{1.13}$$

Thus (1.13) is a <u>necessary</u> condition for the existence of such a periodic solution $x = x(t,\mu, c_\mu)$. Note that from (1.11) we have $x(t, 0, c_0) = e^{tA}c_0$, and (1.13) just expresses the necessary and sufficient condition that $x(t, 0, c_0)$ be a periodic solution of (1.1) with period 2π.

From (1.7) and (1.8) we have

$$e^{2\pi A_j} - E_{\alpha_j} = \begin{pmatrix} 0_2 & 0_2 & \cdots & & 0_2 \\ 2\pi E_2 & 0_2 & & & \vdots \\ \frac{(2\pi)^2}{2!} E_2 & 2\pi E_2 & 0_2 & & \vdots \\ \vdots & \ddots & \ddots & \ddots & \vdots \\ \frac{(2\pi)^{p_j-1}}{(p_j-1)!} E_2 & \cdots & \frac{(2\pi)^2}{2!} E_2 & 2\pi E_2 & 0_2 \end{pmatrix}, \tag{1.14}$$

and from (1.9)

$$e^{2\pi B_j} - E_{\beta_j} = \begin{pmatrix} 0 & 0 & \cdots & & 0 \\ 2\pi & 0 & & & \vdots \\ \frac{(2\pi)^2}{2!} & 2\pi & \ddots & & \vdots \\ \vdots & \ddots & \ddots & \ddots & \vdots \\ \frac{(2\pi)^{\beta_j-1}}{(\beta_j-1)!} & \cdots & \frac{(2\pi)^2}{2!} & 2\pi & 0 \end{pmatrix}. \tag{1.15}$$

From (1.14), (1.15), and (1.10) it is now clear that (1.13) implies that all components c_{0i} of c_0 are zero except possibly those with index i corresponding to the last two rows of any A_j or to the last row of any B_j. These indices are all those with the following forms

$$(1.16)\qquad \begin{aligned} &i = \alpha_1 + \alpha_2 + \dots + \alpha_j - 1 ,\\ &i = \alpha_1 + \alpha_2 + \dots + \alpha_j ,\\ &j = 1, \dots, k ,\\ &i = \alpha_1 + \dots + \alpha_k + \beta_1 + \dots + \beta_j ,\\ &j = 1, \dots, m. \end{aligned}$$

The indices i having the forms given in (1.16) will be called <u>exceptional indices</u>, and the corresponding components of any vector will be called <u>exceptional components</u>. They are $2k + m$ in number, and the exceptional components c_{oi} of c_o are not determined by (1.13).

To proceed further, consider the components of the vector on the left in (1.12) with indices

$$(1.17)\qquad \begin{aligned} &j = 1, 2, \alpha_1 + 1, \alpha_1 + 2, \dots, \alpha_1 + \alpha_2 + \dots + \alpha_{k-1} + 1 ,\\ &\alpha_1 + \alpha_2 + \dots + \alpha_{k-1} + 2, \ \alpha_1 + \dots + \alpha_k + 1,\\ &\alpha_1 + \dots + \alpha_k + \beta_1 + 1, \dots, \alpha_1 + \dots + \alpha_k + \beta_1 + \dots + \beta_{m-1} + 1. \end{aligned}$$

These $2k + m$ indices will be called <u>singular indices</u>, and the corresponding components of a vector <u>singular components</u>. The singular components of $(e^{2\pi A}-E)c_\mu$ are all zero. Thus the singular components of the integral in (1.12) must vanish for all μ sufficiently near $\mu = 0$. This gives

$$(1.18)\qquad \left(\int_0^{2\pi} e^{(2\pi-s)A} f(x(s, \mu, c), s, \mu) ds \right)_j = 0$$

for any j in the set (1.17), where $(\ )_j$ represents the j-th component. In particular if $\mu = 0$, then $x(s, 0, c_o) = e^{sA}c_o$, and thus for the singular indices

$$(1.19)\qquad H_j(c_o) = \left(\int_0^{2\pi} e^{(2\pi-s)A} f(e^{sA} c_o, s, 0) ds \right)_j = 0.$$

If the periodicity of $e^{sA}c_o$ and f are used then (1.19) can be replaced by

$$(1.19)'\qquad H_j(c_o) = \left(\int_0^{2\pi} e^{sA} f(e^{-sA}c_o, -s, 0) ds \right)_j = 0.$$

For $\mu = 0$ all components of c_o other than the exceptional ones are zero. Thus (1.19) is a system of $2k + m$ equations in $2k + m$ unknowns, namely the components c_{oi} of c_o with i exceptional. If $x = x(t, \mu, c_\mu)$ is to exist as a periodic solution it is <u>necessary</u> for (1.19) to have a solution. <u>Note that</u> (1.19) <u>can be written out explicitly without solving the nonlinear system</u> (1.2).

Suppose that the system (1.19) has a solution for the exceptional components of c_o, say $c_{oi} = a_i$. Denoting by a the vector with components a_i for the exceptional indices and zero otherwise, it follows that $p(t) = x(t, 0, a)$ is a periodic solution of (1.1) with period 2π.

We are now in a position to state conditions under which there exists a periodic solution of (1.2) for $|\mu|$ small. We make the following assumptions

(A)

(i) A is a real constant matrix with the canonical form (1.3) - (1.5), with at least one characteristic root of the form iN, where N is an integer,

(ii) μ, f are real, and f has period 2π in t,

(iii) a vector a exists which satisfies (1.13) and (1.19) for $c_0 = a$,

(iv) f, $\partial f/\partial x_i$ are continuous in (x, t, μ) for (x, t) in a vicinity V containing the periodic solution $p(t) = e^{tA}a$ of (1.1), $0 \leq t \leq 2\pi$, in its interior, and for $|\mu| < \delta$, for some $\delta > 0$,

(v) the Jacobian of the $2k + m$ H_j of (1.19), j singular, with respect to the $2k + m$ c_{0i}, i exceptional, does not vanish for $c_{0i} = a_i$.

THEOREM 1. Under the assumptions (A) above there exists a unique periodic solution $x = q(t, \mu)$ of (1.2), of period 2π in t, which is continuous in (t, μ) for all t, and $|\mu|$ sufficiently small, and which for $\mu = 0$ reduces to $q(t, 0) = p(t)$.

REMARK. If assumption (A) (v) does not hold, then a more detailed analysis is required which will not be undertaken here.

PROOF. It will be shown that, for sufficiently small $|\mu|$, (1.12) has a unique solution $c = c_\mu$, continuous in μ, and with $c_0 = a$. From this it follows at once that $q(t, \mu) = x(t, \mu, c_\mu)$ is the desired solution. The existence and uniqueness of a solution $x = x(t, \mu, c)$ for $|\mu|$ and $|c - c_0|$ sufficiently small and with $x(0, \mu, c) = c$ follows directly from the assumption (A) (iv).

To show the existence of $c = c_\mu$ the system of equations (1.12) is replaced by the systems $S_1(\mu, c)$ and $S_2(\mu, c)$, where $S_1(\mu, c)$ consists of the components of (1.12) with non-singular indices, while $S_2(\mu, c)$ consists of the equations (1.18) with singular indices. As discussed after (1.13), $S_1(0, c)$ is a homogeneous linear system for the nonexceptional components c_i of c, and since the determinant Δ of the coefficients is non-vanishing, the only solution is $c_i = 0$, i nonexceptional. $S_2(0, c)$ is just the system of equations (1.19) with $c = c_0$, and by assumption this has a solution $c_{0i} = a_i$, i exceptional. The first partial derivatives of the left side of $S_1(\mu, c)$ with respect to the exceptional components of c are all zero at $\mu = 0$, $c = a$. Thus the overall Jacobian $D(\mu, c)$ of the left sides of $S_1(\mu, c)$ and $S_2(\mu, c)$ with respect to the components of c, when evaluated at $\mu = 0$, $c = a$, is the Jacobian $D_1(\mu, c)$ of the left side of $S_1(\mu, c)$ with respect to the nonexceptional components of c, evaluated at $\mu = 0$, $c = a$, multiplied by the Jacobian $D_2(\mu, c)$ of the left side of (1.18) with respect to the exceptional components of c, evaluated at $\mu = 0$, $c = a$. But $D_1(0, a)$ is just the determinant Δ which is not zero, and $D_2(0, a)$ is just the Jacobian in assumption (A)(v), which is not zero. Therefore, by the implicit function theorem, the combined system $S_1(\mu, c)$, $S_2(\mu, c)$ has a unique solution $c = c_\mu$ for sufficiently small $|\mu|$, which is continuous in μ, and such that $c_0 = a$. This completes the proof.

1.2. The analytic case.

The situation as regards analytic perturbations f is as follows.

THEOREM 2. Suppose the assumptions (A) are satisfied, and in addition let f be an analytic function of (x,μ) for (x, t) in V and $|\mu| < \delta$. Then q is analytic in μ for sufficiently small $|\mu|$.

PROOF. For (μ, c) sufficiently close to $(0, c_0)$, the solution $x = x(t,\mu, c)$ is analytic in μ and c by the existence theorem for such systems. Also, c_μ is analytic in μ by the implicit function theorem for analytic systems. Thus, $q(t,\mu) = x(t,\mu, c_\mu)$ is analytic in μ.

From the practical point of view it is important to know (in the analytic case) whether the periodic coefficients (with period 2π) $q^{(i)}(t)$ in the convergent power series expansion

(1.20) $$q(t,\mu) = \sum_{i=0}^{\infty} \mu^i q^{(i)}(t)$$

can be calculated recursively. As is to be expected this is indeed true. Let the j-th component of $q^{(i)}$ be denoted by $q_j^{(i)}$. Placing (1.20) into (1.2) and comparing coefficients of powers of μ there results

(1.21)
$$q^{(0)}(t) = e^{tA}a,$$
$$\frac{dq^{(1)}}{dt}(t) = A\, q^{(1)}(t) + f(q^{(0)}(t), t, 0),$$
$$\frac{dq^{(2)}}{dt}(t) = A\, q^{(2)}(t) + \sum_{i=1}^{n} \frac{\partial f}{\partial x_i}(q^{(0)}(t), t, 0)\, q_i^{(1)}(t) + \frac{\partial f}{\partial \mu}(q^{(0)}(t), t, 0),$$
$$\frac{dq^{(j)}}{dt}(t) = A\, q^{(j)}(t) + \sum_{i=1}^{n} \frac{\partial f}{\partial x_i}(q^{(0)}(t), t, 0)\, q_i^{(j-1)}(t) + F^{(j)}(t)$$

where $F^{(j)}(t)$ depends only on $q^{(\ell)}(t)$ for $0 \leq \ell \leq j-2$.

That there exist solutions $q^{(i)}$ to the system of differential equations (1.21) follows from the existence of q. It will be shown that each equation in (1.21) has at most one solution, and thus the formal process of solving for the $q^{(i)}$ recursively yields q. Clearly $q^{(1)}$ is determined by the second equation in

(1.21) only to within a periodic solution of the homogeneous equation (1.1). However the requirement that the next equation of (1.21) have a periodic solution $q^{(2)}$ determines $q^{(1)}$ uniquely. For suppose this is not the case. Then there are two distinct solutions for $q^{(1)}$ each of which in the next equation allows for the solution of a periodic $q^{(2)}$. Denoting the differences between the two $q^{(1)}$'s by $\tilde{q}^{(1)}$ and the two $q^{(2)}$'s by $\tilde{q}^{(2)}$, it follows by subtracting the two equations for the $q^{(1)}$'s, from each other and the two for the $q^{(2)}$'s, that

$$\frac{d\tilde{q}^{(1)}}{dt}(t) = A\,\tilde{q}^{(1)}(t)\,, \tag{1.22}$$

$$\frac{d\tilde{q}^{(2)}}{dt}(t) = A\,\tilde{q}^{(2)}(t) + \sum_{i=1}^{n} \frac{\partial f}{\partial x_i}(e^{tA}a,\ t,\ 0)\tilde{q}_i^{(1)}(t). \tag{1.23}$$

If $\tilde{q}^{(2)}(0) = \tilde{a}^{(2)}$, then from (1.23),

$$\tilde{q}^{(2)}(t) = e^{tA}\tilde{a}^{(2)} + \int_0^t e^{(t-s)A} \sum_{i=1}^{n} \frac{\partial f}{\partial x_i}(e^{sA}a,\ s,\ 0)\tilde{q}_i^{(1)}(s)ds.$$

Since $\tilde{q}^{(2)}$ has period 2π this yields

$$(e^{2\pi A}-E)\,\tilde{a}^{(2)} + \int_0^{2\pi} e^{(2\pi - s)A} \sum_{i=1}^{n} \frac{\partial f}{\partial x_i}(e^{sA}a,\ s,\ 0)\tilde{q}_i^{(1)}(s)ds = 0.$$

Taking the components of the above with singular indices, j, it follows that

$$\left(\int_0^{2\pi} e^{(2\pi - s)A} \sum_{i=1}^{n} \frac{\partial f}{\partial x_i}(e^{sA}a,\ s,\ 0)\tilde{q}_i^{(1)}(s)ds\right)_j = 0. \tag{1.24}$$

Clearly $\tilde{q}^{(1)}$, being a periodic solution of (1.22), is of the form $\tilde{q}^{(1)}(s) = e^{sA}\tilde{a}^{(1)}$, where the only non-vanishing components $\tilde{a}_j^{(1)}$ of $\tilde{a}^{(1)}$ are those with exceptional indices. Now (1.24) is linear homogeneous in these $\tilde{a}_j^{(1)}$, and the determinant of the coefficients of the left side of (1.24) with respect to the $\tilde{a}_j^{(1)}$ is precisely the Jacobian of the $2k + m$ H_j of Theorem 1, which is assumed not to vanish. Hence $\tilde{a}^{(1)} = 0$, and thus $\tilde{q}^{(1)}(t) \equiv 0$. Precisely the same argument shows that if the $q^{(\ell)}$ are uniquely determined for $\ell \leqq j$, where $j > 1$, then $q^{(j+1)}$ is also uniquely determined.

THEOREM 3. If the assumptions of Theorem 2 hold, then the analytic solution q of (1.2) can be obtained recursively by solving the system (1.21) for the periodic coefficients $q^{(i)}$, of period 2π, in the convergent power series expansion (1.20). Each $q^{(i)}$, $i \geq 1$, is determined uniquely by the i-th equation in (1.21) and the fact that the $(i + 1)$-st equation has a periodic solution.

1.3. Examples.

As a first example consider the second order differential equation

$$u'' + u = \mu f(u,\ u',\ \mu) + \mu F \cos t, \tag{1.25}$$

where u, μ f, F, are real, and $|\mu|$ is small (Remark: In applications [2] the periodic term $\cos t$ is often replaced by $\cos \omega t$ where ω is near 1. However a change of scale of t has the effect of replacing ωt by t and replacing u by u/ω^2 on the left side of (1.25). If $1/\omega^2 = 1 - \mu\sigma$ then the effect of varying ω can be observed by varying σ. Thus if $\tilde{t} = \omega t$, and $\cdot = d/d\tilde{t}$, (1.25) becomes

$$\ddot{u} + u = \mu[(1 - \mu\sigma) f(u, (1 - \sigma\mu)^{-1/2} \dot{u}, \mu) + \sigma u + (1 - \mu\sigma) F \cos \tilde{t}].$$

Putting $\mu = 0$ in the bracket on the right the bracket becomes $f(u, \dot{u}, 0) + \sigma u + F \cos \tilde{t}$. This is all that enters the H_j terms. Thus in these terms the effect of ω is equivalent to modifying f by adding σu to it. This same device can be used in more general cases). The equation (1.25) can be replaced by a pair of first order equations by setting $u' = v$, but this is not necessary. In any case here the periodicity requirement on a solution of (1.25) gives rise to two equations. Both are singular and both components of c_μ are exceptional. Instead of assuming for $\mu = 0$, $u = c_{01} \cos t + c_{02} \sin t$ it is more convenient to assume $u = c \cos (t + d)$. It is also convenient to translate t in (1.25) so that $u = c \cos t$, while the undetermined d comes into the last term of (1.25) as $F \cos (t - d)$. Using the variation of constants formula on (1.25) the periodicity condition becomes

$$(1.26) \qquad \int_0^{2\pi} \sin s \, [f(c \cos s, - c \sin s, 0) + F \cos(s - d)] ds = 0,$$

$$(1.27) \qquad \int_0^{2\pi} \cos s \, [f(c \cos s, - c \sin s, 0) + F \cos(s - d)] ds = 0,$$

where c and d are to be determined. These are of course the H_j relations for this case. If f is an even function of u', then (1.26) becomes

$$F \int_0^{2\pi} \sin s \cos(s - d) ds = 0 ,$$

which implies $d = 0$ or $d = \pi$. The equation (1.27) then involves c only. A similar simplification occurs if f is odd in u'.

For cases like $f = au + bu^3$ the equation (1.25) is of great practical interest; it is of especial importance to know the dependence of c on a, b, and F.

A more general example is the system (1.2) with A having $\pm i$ as simple characteristic roots and where there are no other characteristic roots of the form iN, N an integer (including $N = 0$). Let the system be written as

$$(1.28) \qquad y' = Sy + \mu g(y, t, \mu),$$

where the constant matrix S is not necessarily in canonical form. Since $\pm i$ are characteristic roots, there exists a constant non-singular real matrix P such that $PS = AP$ where A has the form

(1.29) $$A = \begin{pmatrix} 0 & -1 & \vdots & \\ & & \vdots & 0 \\ 1 & 0 & \vdots & \\ \cdots & \cdots & \cdots & \cdots \\ 0 & & \vdots & A_0 \end{pmatrix}$$

Here A_0 is not necessarily canonical, but it is real and has no characteristic roots of the form iN, for any integer N. If (1.28) is put into the A form by setting $x = Py$ then there results

$$x' = A x + \mu P\, g(P^{-1}x,\ t, \mu),$$

and the H_j relations (1.19)' become

$$H_j(c_0) = \left(\int_0^{2\pi} e^{sA} P\, g(P^{-1}\ e^{-sA} c_0,\ - s,\ 0) ds \right)_j = 0,\ \ j = 1,\ 2,$$

or

(1.30) $$H_j(c_0) = \left(P \int_0^{2\pi} e^{sS}\ g(\ e^{-sS} P^{-1} c_0,\ - s,\ 0) ds\right)_j = 0,\quad j = 1,\ 2.$$

If $p^{(1)}$ and $p^{(2)}$ are the first and second rows of P respectively, then clearly

(1.31) $$H_j(c_0) = p^{(j)} \int_0^{2\pi} e^{sS}\ g(\ e^{-sS} P^{-1} c_0,\ - s,\ 0) ds = 0,\quad j = 1,\ 2.$$

Now $e^{sS}P^{-1}c_0$ represents the general periodic solution of $y' = Sy$. This is associated with the roots $\pm i$. If $y^{(1)}$ and $y^{(2)}$ are two such real, independent, solutions, then $e^{-sS}P^{-1}c_0$ can be replaced by $c_{01}\, y^{(1)}(-s) + c_{02} y^{(2)}(-s)$, and hence (1.31) becomes a pair of equations for c_{01} and c_{02}. If $c_{01} = a_1$, $c_{02} = a_2$ is a solution, and if the Jacobian $\partial(H_1, H_2)/\ \partial(c_{01}, c_{02})$ does not vanish at $c_{01} = a_1$, $c_{02} = a_2$, then Theorem 1 may be applied (provided of course that the other assumptions of (A) hold) to prove the existence of a unique periodic solution $y = q(t, \mu)$ for small enough $|\mu|$.

Note that (1.31) involves only the first two rows of the matrix P. The fact that $PS = AP$ yields

$$\sum_{j=1}^{n} p_{1j} s_{jk} = -p_{2k},\quad \sum_{j=1}^{n} p_{2j} s_{jk} = p_{1k},$$

if $P = (p_{ij})$ and $S = (s_{jk})$. Thus

$$\sum_{j=1}^{n} (p_{1j} + i\ p_{2j})\ s_{jk} = i\ (p_{1k} + ip_{2k}),$$

and hence $(p^{(1)} + ip^{(2)})S = i\ (p^{(1)} + ip^{(2)})$. This implies $u = (p^{(1)} + ip^{(2)})e^{it}$ is a solution of $u' = uS$, and this last property determines $p^{(1)} + ip^{(2)}$ to within a constant factor. Of course, from the practical point of view, one still has to calculate e^{sS}, and it is a question whether this is handled more easily by first reducing S to canonical form or not.

2. PERTURBATION OF AN AUTONOMOUS SYSTEM WITH A VANISHING JACOBIAN

Here the real system

$$x' = Ax + \mu f(x, \mu) \tag{2.1}$$

will be considered where A is a real constant matrix, $|\mu|$ is small, and $f(x, \mu)$ is real continuous in (x, μ) for small $|\mu|$ and x in a region V to be described later. In fact it will be assumed that the $\partial f/\partial x_i$, $i = 1, \ldots, n$, are continuous in (x, μ) for x in V and $|\mu|$ small. With $\mu = 0$ the system (2.1) becomes

$$x' = Ax. \tag{2.2}$$

It is assumed that (2.2) has a periodic solution of period 2π. Note that 2π need not be its least period. It is assumed to be the case that $e^{2\pi A}$ does not have 1 as a simple characteristic root. This last statement is equivalent to the vanishing of the relevant Jacobian J_1.

As in Section 1 it can be assumed with no restriction that A is in real canonical form given by formulas (1.3) - (1.5). Thus (1.6) - (1.10) also hold. However as before C need not be in canonical form.

2.1 <u>Existence of periodic solutions for small</u> $|\mu|$.

Let (2.1) have a unique solution $x = x(t, \mu, c)$, where $x(0, \mu, c) = c$, which exists for t in some finite interval and is continuous in μ for μ near $\mu = 0$. Then, as before,

$$x(t, \mu, c) = e^{tA}c + \mu \int_0^t e^{(t-s)A} f(x(s, \mu, c), \mu)ds. \tag{2.3}$$

Necessary and sufficient for x to be periodic of period $2\pi + \tau$ is that

$$(e^{(2\pi+\tau)A} - E)c + \mu \int_0^{2\pi+\tau} e^{(2\pi+\tau-s)A} f(x(s, \mu, c), \mu)ds = 0,$$

or

$$(e^{2\pi A} - E)c + e^{2\pi A}(e^{\tau A} - E)c + \mu \int_0^{2\pi+\tau} e^{(2\pi+\tau-s)A} f(x(s, \mu, c), \mu)ds = 0. \tag{2.4}$$

If $x = x(t, \mu, c)$ is periodic of period $2\pi + \tau$ and if $c = c_\mu$ and $\tau = \tau(\mu)$ are continuous for small enough μ, and $\tau(0) = 0$, then it follows, since $x(t, 0, c_0) = e^{tA}c_0$, that

$$(e^{2\pi A} - E)c_0 = 0. \tag{2.5}$$

As in Section 1, this implies that only the components of c_0 with exceptional indices can be different from zero. (As before τ and c need in fact only exist for small $\mu > 0$ with limiting values as $\mu \to 0+$).

We assume here that in the canonical form for A there appears at least one matrix of the type A_j. The exceptional indices associated with A_1 are $\alpha_1 - 1$ and α_1. The component of $e^{tA}c_0$ with index α_1 is $(c_0)_{\alpha_1 - 1} \sin N_1 t + (c_0)_{\alpha_1} \cos N_1 t$, and hence, for any specific choice of $(c_0)_{\alpha_1 - 1}$ and $(c_0)_{\alpha_1}$, this sinusoid vanishes for some value of t and has there a non-vanishing first derivative. By continuity, the component x_{α_1} of $x = x(t, \mu, c_\mu)$ must cross the t-axis at some t also. The system (2.1) is invariant under translations in t. Thus, with no restriction, it can be assumed that x_{α_1} vanishes at $t = 0$. This means that $x_{\alpha_1}(0, \mu, c_\mu) = (c_\mu)_{\alpha_1} = 0$ for sufficiently small $|\mu|$, including $\mu = 0$. In fixing this component of x we are then free to determine the period $\tau = \tau(\mu)$. Thus the problem becomes one in obtaining sufficient conditions for the existence of $c = c_\mu$ and $\tau = \tau(\mu)$ (with $(c_\mu)_{\alpha_1} = 0$) satisfying (2.4).

Now, returning to (2.4), we see as in Section 1 that the components of $(e^{2\pi A}-E)c_\mu$ with singular indices are all zero. Thus for these indices

$$(2.6)\qquad e^{2\pi A}\left(\frac{e^{\tau A}-E}{\tau}\right) c_\mu\left(\frac{\tau}{\mu}\right) + \int_0^{2\pi+\tau} e^{(2\pi+\tau-s)A} f(x(s, \mu, c_\mu), \mu)ds = 0.$$

Letting $\mu \to 0$ in (2.6) the term involving the integral tends to a limit, and hence the other term approaches a limit. Since $\tau \to 0$ as $\mu \to 0$, (2.6) gives for the components with singular indices

$$(2.7)\qquad e^{2\pi A} A\, c_0 \lim_{\mu\to 0} \left(\frac{\tau}{\mu}\right) + \int_0^{2\pi} e^{(2\pi-s)A} f(e^{sA}c_0, 0)ds = 0.$$

Changing the variables from s to $2\pi - s$ and using the periodicity of $e^{sA}c_0$ there follows

$$e^{2\pi A} A\, c_0 \lim_{\mu\to 0} \left(\frac{\tau}{\mu}\right) + \int_0^{2\pi} e^{sA} f(e^{-sA}c_0, 0)ds = 0,$$

which can be used instead of (2.7). If at least one singular component of $e^{2\pi A}Ac_0$ is different from zero then (2.7) implies the existence of the limit of τ/μ as $\mu \to 0$, whereas the existence of this limit is not implied by (2.7) if all components of $e^{2\pi A}Ac_0$ with singular indices are zero. The system (2.7), which holds for singular indices only, can be regarded as a system of $2k + m$ equations for the $2k + m$ unknowns consisting of the components of c_0 with exceptional indices ($(c_0)_{\alpha_1} = 0$) and the unknown $\lim (\tau/\mu)$, $\mu \to 0$.

Let

$$(2.8)\qquad H_j(c_0, \nu) = \left(\nu e^{2\pi A} A\, c_0 + \int_0^{2\pi} e^{(2\pi-s)A} f(e^{sA} c_0, 0)ds\right)_j,$$

where j goes through the singular indices. (It is to be observed that H_j is given explicitly in (2.8) as a function of c_0 and ν, and does not require that (2.1) be solved).

For the following existence theorem we <u>assume</u>

(A)

(i) A is a real constant matrix with canonical form (1.3) - (1.5),

(ii) μ, $f = f(x, \mu)$ are real,

(iii) a vector a, with $a_{\alpha_1} = 0$ and with non-exceptional components zero, and a number ν_0 exist which satisfy $H_j(a, \nu_0) = 0$, j singular,

(iv) f, $\partial f/\partial x_i$ are continuous in (x, μ) for x in a vicinity V containing the periodic solution $e^{tA}a$, of period 2π, of (2.2), and for $|\mu| < \delta$, for some $\delta > 0$,

(v) the Jacobian determinant of the $2k + m$ H_j of (2.8) with respect to the $2k + m - 1$ $(c_o)_i$, i exceptional $(i \neq \alpha_1)$, and with respect to ν, does not vanish for $(c_o)_i = a_i$ $((c_o)_{\alpha_1} = 0)$, and $\nu = \nu_o$.

THEOREM 4. Under the assumptions (A) above there exists a periodic solution $x = q(t, \mu)$ of (2.1) with period $2\pi + \tau(\mu)$, where q is continuous in (t, μ) for all t and $|\mu|$ sufficiently small, $\tau = \tau(\mu)$ is continuous in μ, $q(t, 0) = e^{tA}a$, and $\tau(\mu)/\mu \rightarrow \nu_o$, as $\mu \rightarrow 0$. There is no other periodic solution of (2.1) which when $\mu \rightarrow 0$ becomes $e^{tA}a$.

REMARK. If (A) (v) does not hold a more detailed analysis is required which will not be pursued here.

PROOF. Since ν enters H_j linearly the Jacobian will certainly vanish if the coefficients $(e^{2\pi A}Aa)_j = 0$, j singular. Note that as in Section 1 the a_j with non-exceptional indices all must vanish. From this it follows that the terms $(e^{2\pi A}Aa)_j$, j singular, can be different from zero only for those j associated with an A_i which has exactly 2 rows and columns, and for no B_i. <u>Thus</u> (A) (v) <u>really implies that there is at least one</u> A_i, <u>say</u> A_1, <u>and that</u> A_1 <u>must have two rows and columns</u>.

The proof follows that of Theorem 1 closely, and hence will be omitted.

2.2 <u>The analytic case</u>.

THEOREM 5. Let $f = f(x, \mu)$ be analytic in (x, μ) for x in V and $|\mu| < \delta$, and suppose the assumptions of Theorem 4 hold. Then the periodic solution q is analytic in (t, μ) for all t and for $|\mu|$ sufficiently small, and τ is analytic in μ for $|\mu|$ sufficiently small .

PROOF. The proof is very much like that of Theorem 2.

In the analytic case the coefficients in the power series expansions for q and τ can be calculated recursively. Here it is convenient to replace t by s where $t = s(1 + \tau/2\pi)$, and let $q(s(1 + \tau/2\pi), \mu) = p(s, \mu)$. Clearly $p(s, \mu)$ is analytic in μ for small $|\mu|$, and periodic in s of period 2π. Therefore there exist expansions

$$p(s, \mu) = \sum_{i=0}^{\infty} \mu^i \, p^{(i)}(s), \tag{2.9}$$

and

$$(2.10) \qquad \tau/2\pi = \sum_{i=1}^{\infty} \mu^i b_i,$$

where both series converge for small enough $|\mu|$. Clearly, $2\pi b_1 = \nu_0$, and $p^{(0)}(s) = e^{sA}a$. Since the component of $q(0, \mu)$ with index $\alpha_1 = 2$ vanishes, it follows that this component of $p^{(i)}(0)$ must vanish for all $i \geq 0$.

The differential equation (2.1) becomes

$$(2.11) \qquad \frac{dx}{ds} = (1 + \frac{\tau}{2\pi})(Ax + \mu f(x, \mu)).$$

If (2.9) and (2.10) are substituted into (2.11) and coefficients of the powers of μ are compared there results the following system of equations:

$$(2.12) \quad \begin{cases} \dfrac{dp^{(0)}}{ds} = A\,p^{(0)}, \\ \dfrac{dp^{(1)}}{ds} = A\,p^{(1)} + b_1 A\,p^{(0)} + f(p^{(0)}, 0), \\ \dfrac{dp^{(2)}}{ds} = A\,p^{(2)} + b_1 A\,p^{(1)} + b_2 A\,p^{(0)} + b_1 f(p^{(0)}, 0) \\ \quad \vdots \qquad + \displaystyle\sum_{i=1}^{n} \frac{\partial f}{\partial x_i}(p^{(0)}, 0)\, p_i^{(1)} + \frac{\partial f}{\partial \mu}(p^{(0)}, 0) \quad , \\ \dfrac{dp^{(j)}}{ds} = A\,p^{(j)} + b_1 A\,p^{(j-1)} + b_j A\,p^{(0)} + \displaystyle\sum_{i=1}^{n} \frac{\partial f}{\partial x_i}(p^{(0)}, 0)\, p_i^{(j-1)} \\ \quad \vdots \qquad + F^{(j)}, \end{cases}$$

where $F^{(j)}$ depends on $p^{(0)}, p^{(1)}, \ldots, p^{(j-2)}$, and $b_1, b_2, \ldots, b_{j-1}$. That the system (2.12) has a solution for $p^{(i)}$ and b_i is clear.

THEOREM 6. Under the assumptions of Theorem 5 the analytic solution q of (2.1) can be obtained by solving equations (2.12) in succession for the periodic coefficients $p^{(i)}$, of period 2π, of the power series (2.9) for $p(s, \mu) = q(s(1 + \tau/2\pi), \mu)$, and the constants b_i in the expansion (2.10) for $\tau/2\pi$. The $p^{(i)}$ and b_i are uniquely determined in (2.12) by the requirements that $p^{(0)}(s) = e^{sA}a$, $2\pi b_1 = \nu_0$, $p^{(i)}(s + 2\pi) = p^{(i)}(s)$, and $p_2^{(i)}(0) = 0$.

PROOF. Suppose there are two functions $p^{(1)}, \hat{p}^{(1)}$, satisfying the second equation in (2.12), and two constants $b_2, \hat{b}_2$, such that to the pairs $(p^{(1)}, b_2)$, $(\hat{p}^{(1)}, \hat{b}_2)$ there correspond $p^{(2)}, \hat{p}^{(2)}$, respectively satisfying the third equation of (2.12). Subtracting the third equation for one case from that for the other case, and denoting $\tilde{p}^{(1)} = p^{(1)} - \hat{p}^{(1)}$, $\tilde{p}^{(2)} = p^{(2)} - \hat{p}^{(2)}$, $\tilde{b}_2 = b_2 - \hat{b}_2$, we have

$$\frac{d\tilde{p}^{(2)}}{ds} = A\,\tilde{p}^{(2)} + b_1 A\,\tilde{p}^{(1)} + \tilde{b}_2 A\,p^{(0)} + \sum_{i=1}^{n} \frac{\partial f}{\partial x_i}(p^{(0)}, 0)\,\tilde{p}_i^{(1)}.$$

From the second equation for each case follows

$$\frac{d\tilde{p}^{(1)}}{ds} = A\,\tilde{p}^{(1)}.$$

Let $\tilde{p}^{(1)}(0) = \tilde{a}^{(1)}$, and $\tilde{p}^{(2)}(0) = \tilde{a}^{(2)}$. The second (α_1) component of each is zero. Since $\tilde{p}^{(1)}$ is periodic it follows that

$$\tilde{p}^{(1)}(s) = e^{sA}\tilde{a}^{(1)},$$

where only the exceptional components of $\tilde{a}^{(1)}$ can be different from zero. Thus $\tilde{a}^{(1)}$ has at most $2k + m - 1$ components that are not known to be zero. From the differential equation for $\tilde{p}^{(2)}$, and the fact that $\tilde{p}^{(2)}(0) = \tilde{p}^{(2)}(2\pi)$, it follows that

$$(E - e^{2\pi A})\,\tilde{a}^{(2)} = b_1 A \int_0^{2\pi} e^{(2\pi-\sigma)A}\tilde{p}^{(1)}(\sigma)d\sigma$$

$$+ \tilde{b}_2 A \int_0^{2\pi} e^{(2\pi-\sigma)A} p^{(0)}(\sigma)d\sigma$$

$$+ \int_0^{2\pi} e^{(2\pi-\sigma)A} \sum_{i=1}^{n} \frac{\partial f}{\partial x_i}(p^{(0)}(\sigma), 0)\, \tilde{p}_i^{(1)}(\sigma)d\sigma .$$

Since the singular components of the left side vanish the same is true for the right side. Setting $p^{(0)}(\sigma) = e^{\sigma A}a$, and $2\pi b_1 = \nu_0$, it follows that

$$(\nu_0\, e^{2\pi A} A\, \tilde{a}^{(1)} + 2\pi \tilde{b}_2\, e^{2\pi A} A\, a$$

$$+ \int_0^{2\pi} e^{(2\pi-\sigma)A} \sum_{i=1}^{n} \frac{\partial f}{\partial x_i}(e^{\sigma A}a, 0)(e^{\sigma A}\tilde{a}^{(1)})_i\, d\sigma)_j = 0$$

for all singular j. These $2k + m$ equations are linear homogeneous in the $2k + m$ terms consisting of $2\pi\tilde{b}_2$ and the $2k + m - 1$ components of $\tilde{a}^{(1)}$ not fixed at zero. However the determinant made up of the coefficients of these $2k + m$ terms is precisely the Jacobian of the $2k + m$ $H_j(c_0, \nu)$ with respect to (c_0, ν) evaluated at $\nu = \nu_0$, $c_0 = a$. This Jacobian is not zero. Thus $\tilde{b}_2 = 0$ and $\tilde{a}^{(1)} = 0$, which proves that b_2 and $p^{(1)}$ are uniquely determined by (2.12). In the same way it follows that if $p^{(0)}, \ldots, p^{(j-1)}$ and $b_1, \ldots, b_j$ $(j > 2)$ are determined uniquely, then (2.12) determines $p^{(j)}$ and b_{j+1} uniquely, thus yielding the result by induction.

2.3 Examples.

Examples analogous to those in (1.3) can be considered. The case

$$u'' + u = \mu f(u, u', \mu),$$

where u, μ are scalars, yields for the periodicity equations (2.6)

$$c_\mu \frac{(\cos\tau - 1)}{\tau} \cdot \frac{\tau}{\mu} + \int_0^{2\pi+\tau} \sin(\tau - s)\, f(u(s,\mu), u'(s,\mu), \mu)ds = 0$$

$$- c_\mu \frac{\sin\tau}{\tau} \cdot \frac{\tau}{\mu} + \int_0^{2\pi+\tau} \cos(\tau - s)\, f(u(s,\mu), u'(s,\mu), \mu)ds = 0,$$

where c_μ here represents a scalar. Letting $\mu \to 0$ and recalling that $\tau(0) = 0$ there results

$$\int_0^{2\pi} \sin s\; f(c_0 \cos s, - c_0 \sin s, 0)ds = 0,$$

$$- c_0\, \nu + \int_0^{2\pi} \cos s\; f(c_0 \cos s, - c_0 \sin s, 0)ds = 0.$$

If the first equation has a solution $c_0 = a_0 \neq 0$, then the second equation determines a solution for ν.

The analogue of the more general example in (1.3) can also be considered. This example is the subject of a paper by Bulgakov [1]. In the terminology of (1.28) the equation here is

$$y' = Sy + \mu g(y, \mu), \tag{2.14}$$

where S satisfies the same conditions as stipulated after (1.28). If $y^{(1)}$ is a solution of period 2π of $y' = Sy$, then analogous to (1.30) we have

$$(\nu e^{2\pi A} A c_0 + \int_0^{2\pi} e^{sA} P g(P^{-1} e^{-sA} c_0, 0) ds)_j = 0, \quad j = 1, 2,$$

or

$$H_j(c, \nu) = c \nu P S y^{(1)}(2\pi) + \int_0^{2\pi} P e^{sS} g(c y^{(1)}(-s), 0) ds)_j = 0, \quad j = 1, 2, \tag{2.15}$$

where c and ν are numbers to be determined. Since $S y^{(1)} = dy^{(1)}/ds$ equations (2.15) may be put in the form

$$H_j(c, \nu) = p^{(j)} (c \nu \frac{dy^{(1)}}{ds}(2\pi) + \int_0^{2\pi} e^{sS} g(cy^{(1)}(-s), 0) ds) = 0, \quad j = 1, 2. \tag{2.16}$$

If the pair of equations (2.16) have a solution $c = a$, $\nu = \nu_0$, and if the Jacobian $\partial(H_1, H_2)/\partial(c, v)$ is not zero at $c = a$, $\nu = \nu_0$, then (2.14) has a periodic solution $y = q(t, \mu)$ of period $2\pi + \tau(\mu)$ (provided of course the other assumptions of (A) hold). Because (2.16) is linear in ν, ν can be eliminated resulting in a single equation for c.

BIBLIOGRAPHY

[1] BULGAKOV, B. V., "Periodic Processes in Free Pseudo-Linear Oscillatory Systems," Journal of the Franklin Institute, Vol. 235, (1943).

[2] STOKER, J. J., "Nonlinear Vibrations," New York, (1950).

DYNAMICAL SYSTEMS WITH STABLE STRUCTURES

By H. F. DeBaggis

In the study of nonlinear problems it is difficult for the mathematician to find rich classifications of nonlinear systems which are sufficiently homogeneous in their properties to yield an interesting theory. Strong analytic conditions (restricting, for example, the functions to polynomials) will often still admit systems with pathological behaviors. For physical systems to perform certain operations they must, if they are to be useful, possess a certain degree of stability. Small perturbations should not affect the essential features of the system. Since the physical components of a system can never be duplicated exactly, experimental verification would be impossible unless the system remain stable under small variations. The stability requirements of experiments provide a clue to the restrictions a mathematician should place on his nonlinear problems.

A. Andronov and L. Pontrjagin [2] attempted a mathematical description of structurally stable systems (systemes grossier). The "essential features" of systems to be preserved under small perturbations are described mathematically by the topological character of the phase portrait. They gave a statement of the main results of the theory of structural stability. And Andronov [1] showed that only stable limit-cycles can represent real self-oscillating phenomena in such systems. The results of the theory of structural stability were restated and used by A. Andronov and C. Chaikin [1] but again proofs were omitted.

In this paper we give a complete treatment of the theory of structural stability. We have relaxed the conditions of analyticity which were imposed on the functions by [2] and merely require that they have continuous first partials. Furthermore, though we consider a region bounded by a simple closed curve, from our methods of proof, it will be obvious that these results hold for regions bounded by any number of simple closed curves of the type under consideration. In conclusion we add some remarks concerning the structural stability of systems with respect to small variations of the parameter. A more detailed discussion of these concluding remarks will appear elsewhere.

* The author wishes to express his gratitude to Professor Lefschetz for the opportunity to make this study and for his constant encouragement and advice. Thanks are also due Professor M. M. Schiffer for his many helpful suggestions.

PART I

In the first part of this paper we shall show that structural stability imposes certain necessary conditions on the nature of the critical points, separatrices and limit-cycles which a dynamical system may possess.

We deal with a dynamical system

$$\text{(A)} \qquad \frac{dx_i}{dt} = P_i\,(x_1, x_2) \qquad (i = 1, 2),$$

where x_1, x_2 are Cartesian coordinates in the plane and the functions P_i $(i = 1, 2)$ have continuous first partial derivatives within and on the boundary of a closed region G. The boundary of G is assumed to be a simple closed curve L with continuously turning tangent.

The system (A) defines a vector field and the vector (P_1, P_2) is referred to as the velocity vector (at the representative point $M(x_1, x_2)$ of the system). A differentiable arc λ is said to be without contact with the vector field whenever the velocity vector is neither zero or tangent to λ at any of its points. We shall assume that these properties hold for every point of the simple closed curve L and call L a cycle without contact. This implies that once a trajectory (a curve tangent to the field) enters [leaves] G it cannot leave [enter] G. Without loss of generality we may consider the trajectories of (A) cutting the cycle without contact L as entering G with increasing time.

In addition to the system (A) we shall consider the perturbed system,

$$\text{(B)} \qquad \frac{dx_i}{dt} = P_i\,(x_1, x_2) + p_i\,(x_1, x_2) \qquad (i = 1, 2)$$

where the functions p_i have continuous first partials for all points $M(x_1, x_2)$ belonging to G.

Definition 1. Let $\rho\,[M_1, M_2]$ denote the Euclidean distance between two points in the plane. The system (A) is said to be structurally stable in G if for each $\varepsilon > 0$, there exists a $\delta(\varepsilon) < 0$ such that for all $p_i(x_1, x_2)$ $(i = 1, 2)$ which belong to the class C^1 and satisfy the condition

$$|p_i| < \delta, \quad \frac{\partial p_i}{\partial x_j} < \delta \quad (i = 1, 2; \; j = 1, 2)$$

there exists a topological transformation T of G into itself with the additional properties:

1) $\rho\,[M\; T(M)] < \varepsilon$, $[M \in G, \; T(M) \in T(G)]$,

2) T maps trajectories of (A) into trajectories of (B).

Thus, if the system (A) is structurally stable, the perturbed system (B) has the same phase portrait qualitatively as the system (A).

It will be convenient to recall here the definition of Ω and A limit sets.

Definition 2. Let γ be a trajectory, $\gamma^+_{M_0}$ the subset of γ consisting of the point M_0 and all the points of γ traversed after M_0, $\gamma^-_{M_0}$ the analogue referring to the points traversed before M_0. We refer to $\gamma^+_{M_0}$ and $\gamma^-_{M_0}$ as the positive and negative half trajectories determined by M_0. The closures $\bar{\gamma}^+_{M_0}$, $\bar{\gamma}^-_{M_0}$ gives rise, through their intersections, to two new sets

$$\Omega(\gamma) = \bigcap_{M \in \gamma^+_{M_0}} \bar{\gamma}^+_{M_0}, \quad A(\gamma) = \bigcap_{M \in \gamma^-_{M_0}} \bar{\gamma}^-_{M_0}$$

the Ω and A limit sets of γ. (Birkhoff).

The points of G for which $(P_1, P_2) = (0, 0)$ are called critical points of (A). These points are characterized topologically as points through which no trajectories pass and which are limit sets. Equivalently, they have been described [Lefschetz [6]] as trajectories which consist of a single point.

For the sake of completeness we include here some well known facts about the behavior of trajectories in the neighborhood of critical points.

In what follows we shall be interested particularly in the critical points (x_1^0, x_2^0) for which the Jacobian

$$\left[\frac{D(P_1, P_2)}{D(x_i, x_j)}\right] (x_1^0, x_2^0) = \Delta(x_1^0, x_2^0) \neq 0.$$

For simplicity, let us assume (x_1^0, x_2^0) is at the origin. Then by Taylor's Theorem

$$P_i(x_1, x_2) = a_{i1}x_1 + a_{i2}x_2 + g_i(x_1, x_2)$$

$$\lim_{\substack{x_1 \to 0 \\ x_2 \to 0}} \frac{g_i(x_1, x_2)}{|x_1 - x_2|} = 0. \qquad (i = 1, 2)$$

Hence $\Delta(0, 0) \neq 0$ implies $\begin{vmatrix} a_{11} & a_{12} \\ a_{21} & a_{22} \end{vmatrix} \neq 0$ and $\frac{\partial P_i}{\partial x_j} = a_{ij}$, $(i, j = 1, 2)$.

It is known [cf. v. g. Hurewicz [5]] that under these conditions the topological behavior of the trajectories g (A) in the neighborhood of the origin is the same as that of the system

$$\frac{d\xi_i}{dt} = a_{i1}\xi_1 + a_{i2}\xi_2 \quad (i = 1, 2).$$

Hence, when $\Delta(0, 0) > 0$ and $\sigma(0, 0) = -\left[\frac{\partial P_1}{\partial x_1} + \frac{\partial P_2}{\partial x_2}\right](0, 0) \neq 0$ the origin is either a node or a focus depending on whether the roots of the characteristic equation $\lambda^2 + \sigma\lambda + \Delta = 0$ are real and of the same sign or the complex conjugate. The trajectories will tend toward [issue away from] the origin if the real parts of the roots are < 0 [> 0].

It is also well known [Hurewicz [5]] that in this case the origin can be enclosed within an ellipse of small radius such that once a trajectory enters the ellipse it cannot leave. That is, an ellipse which is without contact with the vector field. We shall refer to these ellipses as <u>critical curves</u> and to the regions enclosed by them as <u>critical regions</u>.

When $\Delta(0, 0) > 0$ and $\sigma(0, 0) = 0$, the origin may be either a center or a focus depending on the nature of the higher degree terms. We shall refer to a focus of this kind as a <u>weak focus</u>.

LEMMA 1. Let $M_0 = \Omega(\gamma)$ $[A(\gamma)]$ be a node or focus of a structurally stable system (A). Under the transformation T, required by Definition 1, $T(M_0) = \Omega(T(\gamma))$ $[A(T(\gamma)]$ in $T(G)$.

Let the ε required by Definition 1 be so small that we can isolate M_0 by a critical region R of radius $r > \varepsilon$ and such that any other critical point is a distance $> \varepsilon$ from its boundary. This same critical region contains exactly one critical point of each perturbed system (B). But, for δ small, the velocity vectors of (A) and (B) have approximately the same direction on the boundary of R. Therefore all the vectors of (B) will be directed to the interior [exterior] of R and so $T(M_0) = \Omega[T(\gamma)]$ $[A[T(\gamma)]]$.

Since the Poincaré index, which is $+1$ for a node or focus and -1 for a saddle point, is a topological invariant [6], we can state

LEMMA 2. Limit sets which are nodes or foci [saddle points] cannot, under a topological transformation, map into saddle points [nodes or foci].

LEMMA 3. A structurally stable system (A) can have only a finite number of critical points in G.

Since P_i $(i = 1, 2)$ are C^1 in the closed region G

$$P_i = P_i + p_i \quad (i = 1, 2)$$

where the P_i's are polynomials and $|p_i| < \delta$ in G.(1) Thus according to Definition 1, the system

$$\text{(B)} \qquad \dot{x}_i = P_i \quad (i = 1, 2),$$

where the P_i's are relatively prime, is an admissible perturbation. But the P_i's of (B) have only a finite number of common zeros. Now, because of the topological character of the critical points, the transformation required by Definition 1 must map critical points of (A) into critical points of (B). Hence (A) can have at most a finite number of critical points.

(1) P_1, P_2 can be taken as relatively prime since we can always add a small constant to one of them.

THEOREM 1. The critical points of a structurally stable system (A) can only be nodes, foci and saddle points. That is, if (x_1^o, x_2^o) is a critical point of the system, then

a) $$\Delta(x_1^o, x_2^o) \neq 0$$

b) if $\Delta(x_1^o, x_2^o) > 0$, $\sigma(x_1, x_2) = -(\frac{\partial P_1}{\partial x_1} + \frac{\partial P_2}{\partial x_2}) \neq 0.$

In the proof of the theorem we shall assume that a) and b) are not satisfied and derive a contradiction. Without loss of generality, in the proof, we may assume (x_1^o, x_2^o) at the origin. Moreover, since, by Lemma 3, each critical point of (A) is an isolated point we may encircle the origin by a circle of radius $r > \varepsilon$, and choose $\delta(\varepsilon)$ so small that there is exactly one critical point of the perturbed system within this neighborhood.

Suppose first that $\Delta = 0$, i.e. the roots of the characteristic equation $\lambda^2 + \sigma\lambda + \Delta = 0$ are $\lambda_1 = 0$, $\lambda_2 = -\sigma$. The system

(B) $$\begin{aligned} x_1 &= P_1 + a_1x_1 + a_2x_2 \\ x_2 &= P_2 + a_1x_1 + a_2x_2 \end{aligned}$$

where $|a_ix_i| < \delta$, $|a_i| < \delta$ $(i = 1, 2)$ is according to Definition 1, an admissible perturbation.

For a proper choice of the a_i's, the system (B) will have either a node, focus, or saddle point at the origin. For example, if $\sigma > 0$, the system

(B_1) $$\dot{x}_i = P_i + ax_i \quad (i = 1,2;\ o < a < \sigma)$$

has a node at the origin while

(B_2) $$\dot{x}_i = P_i - ax_i \quad (i = 1, 2;\ o < a < \sigma)$$

has a saddle point there; if $\sigma = 0$, (B_1) has a node at the origin and

(B_3) $$\begin{aligned} \dot{x}_1 &= P_1 - ax_2 \\ \dot{x}_2 &= P_2 - ax_1 \end{aligned} \qquad [\text{sign } a = \text{sign } (\frac{\partial P_1(0, 0)}{\partial x_2} + \frac{\partial P_2(0, 0)}{\partial x_1})]$$

has a saddle point there.

But both (B_1), (B_2) $[(B_3)$ if $\sigma = 0]$ are admissible perturbations of (A). Hence there is a topological transformation mapping a node of (B_1) into a saddle point of (B_2) $[(B_3)]$ in contradiction to Lemma 2.

Next we consider the case $\Delta(0, 0) > 0$, $\sigma = 0$.

For $a > 0$, the above systems (B_1), (B_2) have foci, at the origin, which are Ω and A limit sets, respectively. The same reasoning as above leads to the existence of an ε-transformation which maps an Ω limit set of

(B_1) into an Ω limit-set of (B_2) contradicting Lemma 1. This completes the proof of Theorem 1.

THEOREM 2. A separatrix of a structurally stable system (A) cannot issue from [tend toward] a saddle point and terminate [issue from] another saddle point in G.

We prove the theorem by contradiction.

Let γ be a separatrix of (A) issuing from a saddle point (x_1^0, x_2^0) and tending toward another saddle point (x_1', x_2'). Let $\bar{\gamma}$ denote the closure of γ, i.e. $\bar{\gamma}$ contains γ and the points (x_1^0, x_2^0), (x_1', x_2'). We surround $\bar{\gamma}$ by a simple closed curve C which is a distance $> \varepsilon$ from the points of $\bar{\gamma}$. We further require that C contain no critical points other than (x_1^0, x_2^0) (x_1', x_2').

We now consider the system

$$\text{(B)} \qquad \begin{aligned} \dot{x}_1 &= P_1 \cos \delta(\varepsilon) \mp P_2 \sin \delta(\varepsilon) \\ \dot{x}_2 &= P_2 \cos \delta(\varepsilon) \pm P_1 \sin \delta(\varepsilon). \end{aligned}$$

For suitably chosen $\delta(\varepsilon)$, (B) is an admissible comparison system and has no critical points within C other than the saddle points (x_1^0, x_2^0), (x_1', x_2'). The perturbed system (B) simply effects a rotation of magnitude δ on the velocity vectors of (A). Let us assume the rotation to be in the clockwise direction.

Now because of the requirements on the transformation T by Definition 1, $T(\gamma)\varepsilon(B)$ must also issue from (x_1^0, x_2^0) and tend toward (x_1', x_2'). Denote by $T(\gamma_0)$ the separatrix of (B) which is adjacent to $T(\gamma)$ in the clockwise direction and tending toward (x_1^0, x_2^0) and by $T(\gamma_1)$ the separatrix adjacent to $T(\gamma)$ from the counter-clockwise direction and issuing from (x'_1, x_2'). Since there are no critical points of (B) in C, these trajectories must intersect C. Call K that portion of C between these intersections (see Figure 1). Define R to be the region bounded by $\bar{\gamma}$, $T(\gamma_0)$, $T(\gamma_1)$ and k.

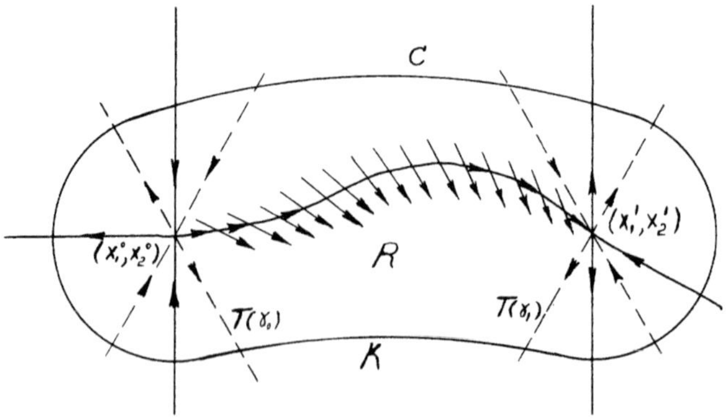

Figure 1

Since the perturbation effects a counter-clockwise rotation on the velocity vectors of (A), $T(\gamma)$ must enter R as it issues from (x_1^0, x_2^0). Furthermore all the trajectories of (B) which cross $\bar{\gamma}$ are entering R. Hence $T(\gamma)$ cannot tend toward (x_1', x_2') unless it first leaves R. But $T(\gamma)$ cannot cross K since the distance from $\bar{\gamma}$ to K is greater than ε. Neither can $T(\gamma)$ intersect $T(\gamma_0)$ and $T(\gamma_1)$ because of the Uniqueness Theorem of differential equations. Therefore $T(\gamma)$ cannot issue from (x_1^0, x_2^0) and tend toward (x_1', x_2'). This completes the proof of the theorem.

By similar reasoning one can readily prove

THEOREM 3. In a structurally stable system a separatrix which is issuing from a saddle point cannot tend toward the same saddle point.

We now make some statements concerning the closed trajectories, i.e., the limit-cycles. They are represented in the form $x_i = f_i(t)$ $(i = 1, 2)$ where the functions f_i are periodic with period τ, say. These trajectories are further characterized by the fact that at least on one side the neighboring trajectories either spiral towards them or away from them. That is to say, if η is a small normal at a point N of a limit cycle $\bar{\gamma}$ and if a trajectory γ intersects η at a point M_0, γ will intersect η again at a point M_1 which lies between M_0 and N if γ spirals towards $\bar{\gamma}$: if γ spirals away from γ, M_0 will lie between M_1 and N. When the trajectories on the exterior and interior of $\bar{\gamma}$ spiral toward [away from] $\bar{\gamma}$, we shall say that is a stable [unstable] limit-cycle.

THEOREM 4. A structurally stable system (A) has only a finite number of limit-cycles in G.

Since under the trajectory preserving ε-mapping required by Definition 1, closed trajectories of (A) must map onto closed trajectories of each admissible perturbed system, to prove our theorem it will suffice to show that the perturbed system

$$\text{(B)} \qquad x_i^0 = P_i(x_1, x_2) \quad (i = 1, 2)$$

has only a finite number of limit-cycles. To this end we shall assume (B) has an infinite number of limit-cycles in G and derive a contradiction. [4].

We recall that since (A) is assumed stucturally stable (B) has only finite number of critical points in G. Hence we can assume, in the proof of our theorem, that there are infinitely many limit-cycles surrounding one critical point M_0 of (B). Through M_0 construct a rectifiable curve k which extends to the boundary of G but which does not pass through any other critical point of (B). Since we have supposed there are infinitely many limit-cycles surrounding M_0, there is on k a limit-point M of the intersections of these cycles with k. Let η be a small normal to the trajectory of (B) which passes through M. On the segment $\bar{\eta} = \overline{M'MM''}$ of η (M between M', M") there pass infinitely many limit-cycles of (B).

It is well known ([3] p. 17) that the trajectory γ passing through M must be one of the following types:

1) a limit-cycle solution of (B);

2) a solution tending toward [issuing from] a limit-cycle, node, or focus;

3) a solution issuing from the boundary of G;

4) a solution tending toward [issuing from] a saddle point.

We shall eliminate each case.

1) γ <u>cannot be a limit-cycle</u>. Let us consider the arc $\bar{\eta}_1 = \overline{M'M}$ of $\bar{\eta}$ and assume this arc contains infinitely many limit-cycles. Then the trajectory Γ_1 passing through the point M_1 of η_1 will intersect $\bar{\eta}_1$ again in a point M_2 where $M_2 = \varphi(t_1 + t_2, M_1)$ $[M_1 = \varphi(t_1, M_1)]$ and φ is an analytic function of t since the solutions of (B) in G are analytic. Under the assumption that γ is a limit of limit-cycles we must have

$$\varphi(t_{i+1}, M_{i+1}) = \varphi(t_i, M_i)$$

infinitely often on the closed interval $\bar{\eta}_1$. Hence by the Uniqueness Theorem

$$\varphi(t_{i+1}, M_{i+1}) \equiv \varphi(t_i, M_i)$$

and so η_1 contains a band of closed solutions in contradiction to our assumption that γ is a limit of limit-cycles. The same reasoning applies to the arc $\eta_2 = \overline{MM''}$ of $\bar{\eta}$.

2) γ <u>cannot be a trajectory of types</u> 2) <u>or</u> 3). In fact if γ were of these types, then by the continuity theorem with respect to the initial conditions there is a neighborhood of M such that all the trajectories passing through this neighborhood are also of types 2) or 3).

3) γ <u>cannot tend toward [issue from] a saddle point</u>. For suppose that γ did tend toward a saddle point. But (B) is by assumption an admissible perturbation of the structurally stable system (A). Hence by Theorems 3 - 4 (B) cannot contain a trajectory which tends towards and issues from a saddle point. This completes the proof of our theorem.

With each limit-cycle $\bar{\gamma}$ we associate the characteristic exponent

$$h(\bar{\gamma}) = \frac{1}{\tau} \int_0^{\tau} \left(\frac{\partial P_1}{\partial x_1} + \frac{\partial P_2}{\partial x_2}\right) dt,$$

where τ is the period of $\bar{\gamma}$. It is well known [1] that when $h(\bar{\gamma}) < 0$ $[> 0]$, $\bar{\gamma}$ is a stable [unstable] limit-cycle, i.e., an Ω [A] limit-set. When $h(\bar{\gamma}) = 0$ no definite statement can be made as to the stability or unstability of $\bar{\gamma}$. We shall show that if (A) is structurally stable, it cannot possess, in G, a limit-cycle $\bar{\gamma}$ such that $h(\bar{\gamma}) = 0$. But first we shall prove the following lemma which is a slight extension of the continuity theorem with respect to the initial conditions.

Let us write the system (A) in the vector form

$$(\bar{A}) \qquad \dot{x} = P(x)$$

where $\dot{x}$, P. denote the column vectors $\begin{pmatrix}\dot{x}_1\\ \dot{x}_2\end{pmatrix}$, $\begin{pmatrix}P_1\\ P_2\end{pmatrix}$, respectively. In this notation the perturbed system (B) becomes

$$(\bar{B}) \qquad \dot{y} = P(y) + p(y).$$

The solutions of $(\bar{A})$, $(\bar{B})$ are given by the equations

$$x = \varphi(t - t_o, M_o) \qquad y = \Psi(t - t_o, \mathfrak{m}_o)$$

where $M_o = \varphi(0, M_o)$, $\mathfrak{m}_o = \Psi(0, \mathfrak{m}_o)$. Now

$$|x - y| \leq \left| \int_{t_o}^{t} [P(x) - P(y)]dt + M_o - \mathfrak{m}_o \right|$$

$$\leq \int_{t_o}^{t} [|P(x) - P(y)| + |p(y)|]dt + |M_o - \mathfrak{m}_o|$$

$$\leq \int_{t_o}^{t} (K\,|x - y| + \quad)dt + |M_o - \mathfrak{m}_o|$$

and by Gronwall's Lemma[2] we obtain

$$|x - y| \leq [\delta\tau + (M_o - M_o)]\, e^{K\tau} \quad (\text{for } t_o \leq t \leq \tau),$$

where $|p| < \delta$ and K is the Lipschitz constant for P in G.

In particular if for a given number $\varepsilon > 0$ and interval of time we choose $\delta < \frac{\varepsilon}{2\tau e^{K\tau}}$, then $|x - y| < \varepsilon$ (for $t_o \leq t \leq \tau$) whenever $|M_o - \quad _o| \ \frac{\varepsilon}{e^{K\tau}}$. Thus we have proved

> LEMMA 4. Given any number $\varepsilon > 0$ and interval of time τ there exists a $\delta(\varepsilon, \tau) > 0$ such that $|x - y| < \varepsilon$ (for $t_o \leq t \leq \tau$) whenever $|M_o - \mathfrak{m}_o| < \delta$.

Let us consider a limit-cycle $\bar{\gamma}$ (in G) which is an Ω set of the structurally stable (A). Let η be a small normal to $\bar{\gamma}$ and call M_o, M_1 the first and second intersections (in the sence of increasing time) of a trajectory γ (exterior to $\bar{\gamma}$) with η. Let M_o', M_1' be the analogues of M_o, M_1 respectively for a trajectory γ' (interior to $\bar{\gamma}$). We denote the closed arcs of η between M_o, M_1 and M_o', M_1' by $\eta(M_oM_1)$, $\eta(M_o'M_1')$ respectively and the analogues for γ, γ' by $\gamma(M_oM_1)$, $\gamma'(M_o'M_1')$ respectively. (see Figure 2).

(2) Gronwall's Lemma: If $z(x)$ is continuous in $x_o - h \leq x \leq x_o + h$ and $z(x) \leq \int_{x_o}^{x_o+h} (Kz + A)dx + B$, where K, A, B are positive constants, then $z(x) \leq (Ah + B)e^{Kh}$.

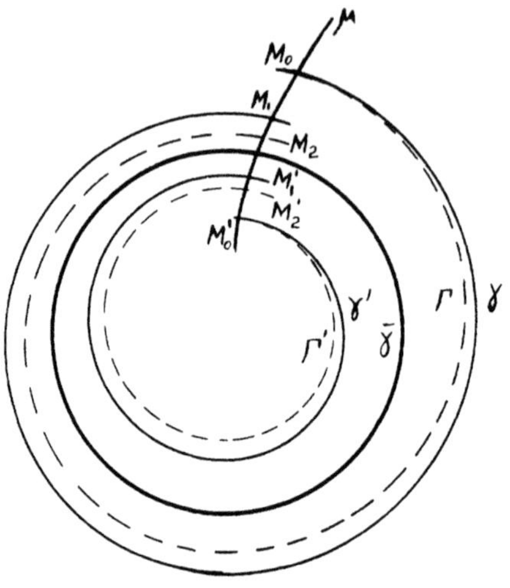

Figure 2

Let us now consider a uniform counter clockwise rotation of the vector field (A) given by

$$(B_1) \qquad \begin{aligned} \dot{x}_1 &= (\cos \delta)P_1 - (\sin \delta)P_2 \\ \dot{x}_2 &= (\cos \delta)P_2 + (\sin \delta)P_1 \end{aligned}$$

By Lemma 4, for δ small enough, the trajectory Γ of (B) which passes through M_0 intersects $\eta_0(M_0M_1)$ again in a neighborhood of M_1 (in a point M_2, say) and Γ does not intersect $\bar{\gamma}$. Similarly, the trajectory Γ' of

$$(B_2) \qquad \begin{aligned} \dot{x}_1 &= (\cos \delta)P_1 + (\sin \delta)P_2 \\ \dot{x}_2 &= (\cos \delta)P_2 - (\sin \delta)P_1 \end{aligned}$$

which passes through M_0' intersects $\eta_0(M_0'M_1')$ again in the point M_2'

Now consider the region R' bounded by $\eta(M_0M_2) \cup \Gamma(M_0M_2)$ and $\eta(M_0'M_2') \cup \Gamma'(M_0'M_2')$. Since Γ and Γ' are trajectories of the rotated fields, all the trajectories of (A) intersecting $\Gamma(M_0M_2)$, $\Gamma'(M_0M_2)$ are entering R'. Since by assumption $\bar{\gamma}$ is an Ω set and since $\eta(M_0M_2)$, $\eta(M_0'M'_2)$ are arcs of the normal η to $\bar{\gamma}$, all the trajectories of (A) which intersect these arcs are also entering R'. Next we smooth out the corners of the two curves bounding R' so as to obtain two smooth curves $\bar{\gamma}_{+\delta}$, $\bar{\gamma}_{-\delta}$ which are without contact with the vector field of (A). The region R bounded by $\bar{\gamma}_{+\delta}$, $\bar{\gamma}_{-\delta}$ is clearly a critical region containing $\bar{\gamma}$.

Now with the same reasoning as in Lemma 1 we can prove

LEMMA 5. *If $\bar{\gamma}$ is a limit-cycle solution of a structurally stable system (A) and $\bar{\gamma}$ is an Ω limit-set, then the corresponding solution of (B) under the transformation T* is also an Ω limit-set.

We next prove

THEOREM 6. If $\bar{\gamma}$ is a limit-cycle solution of a structurally stable system (A), then $h(\bar{\gamma}) \neq 0$.

To prove our theorem it will be sufficient, in view of Lemma 5, to show that if $h(\bar{\gamma}) = 0$, then for an admissible perturbed system (B), $h(T(\bar{\gamma})) \neq 0$ and may have an arbitrary sign.

Let us consider a function $F(x,y)$ of the class C^1 with the property that $F(x, y) = 0$ on $\bar{\gamma}$ and $F_x(x, y) \neq 0$ on $\bar{\gamma}$. An example of such a function is $F(x, y) = \eta(x, y)\ N(x, y)$ where $\eta(x, y)$ is the distance function in a region R_δ bounded by the δ-parallels to $\bar{\gamma}$. Let sign $\eta < 0$ outside of $\bar{\gamma}$ and < 0 inside $\bar{\gamma}$. We define

$$N(x, y) = \begin{cases} e^{-(\tan \frac{\pi n}{\delta})^2} & \text{in } R_\delta (n \text{ the normal distance from } \bar{\gamma}) \\ 0 \text{ in } G - R_\delta \end{cases}$$

The perturbed system

$$\text{(B)} \qquad \begin{aligned} \dot{x}_1 &= P_1(x, y) \pm aF(x, y)F_x \\ \dot{x}_2 &= P_2 \end{aligned}$$

is for $|a|$ sufficiently small an admissible perturbation.

Since $F(x, y) = 0$ on $\bar{\gamma}$, $\bar{\gamma}$ is a solution of (B). Hence $\bar{\gamma} = T(\bar{\gamma})$ and the characteristic exponent for $T(\bar{\gamma})$ is

$$h(t(\bar{\gamma})) = \frac{1}{\tau} \int_0^\tau \left[\frac{\partial P_1}{\partial x_1} + \frac{\partial P_2}{\partial x_2}\right] \pm aF_x^2 \ dt.$$

Now by assumption

$$h(\bar{\gamma}) = \frac{1}{\tau} \int_0^\tau \left[\frac{\partial P_1}{\partial x_1} + \frac{\partial P_2}{\partial x_2}\right] dt = 0.$$

Hence $h(T(\bar{\gamma})) = \frac{\pm a}{\tau} \int_0^\tau F_x^2 \ dt.$

This competes the proof of the theorem.

We shall say that a system (A) satisfies in G the condition (*) if in the interior of the cycle L, assumed to be without contact, the following conditions are fulfilled:

1) The critical points are such that $\Delta \neq 0$, and if $\Delta > 0$, $\sigma = 0$.
2) There are only a finite number of periodic solutions and they are such that $h \neq 0$.
3) There are no separatrices both issuing from a saddle point and tending toward a saddle point.

For brevity, if a system of type (A) satisfies the conditions (*) in G, we shall denote it by (A*)

To simplify our terminology we shall also call critical points trajectories.

In view of the known results of Bendixson [3], a system (A*) can have only the following types of trajectories:

I. States of equilibrium
   1. Nodes, foci $[\Delta > 0, \sigma \neq 0]$
   2. Saddle points $[\Delta < 0]$

II. Limit-cycles
   1. $h \neq 0$ [$h < 0$ strongly stable, $h > 0$ strongly unstable]

III. Separatrices
   1. Issuing from a node, focus, limit-cycle or tending toward a node, focus, limit-cycle.
   2. Entering the region G.

IV. Trajectories whose only limit sets in G are nodes, foci and limit-cycles.
   1. Issuing from a node, focus or limit-cycle and tending toward a node, focus or limit-cycle.

V. Trajectories (not separatrices) entering G.
   1. Tending toward a node, focus or limit-cycle.

PART II

Our aim in this section will be to prove that a dynamical system of types (A*) is structurally stable in G.

<u>Definition 3</u>. Let $M = \varphi(t - t_0, M_0)$ $M_0 = \varphi(0, M_0)$ be the path of a trajectory γ. γ is said to be <u>positively</u> (<u>negatively</u>) <u>Liapounoff</u> stable (briefly, $\mathcal{L}^+[\mathcal{L}^-]$ stable) in G, if for each $M_0 \varepsilon \gamma$, $M = \varphi(t - t_0, M_0)$ satisfies the following two conditions:

1. for all $t > t_0$ $[t < t_0]$ $M \varepsilon G$,
2. for all $\varepsilon > 0$ there exists a $\delta(\varepsilon, M_0)$ such that for all $t > t_0$ $[t > t_0]$ the distance $\rho[\varphi(t - t_0, M_0), \varphi(t - t_0, M_0')] < \varepsilon$ $(M_0' \varepsilon \gamma')$ whenever

$$\rho[M_0, M_0'] < \delta .$$

A trajectory γ is said to be $\mathcal{L}^+[\mathcal{L}^-]$ <u>unstable</u> in G if it satisfies the condition 1) but not 2).

<u>Definition 4</u>. A trajectory is called <u>singular</u> if it is either $\mathcal{L}^+$ or $\mathcal{L}^-$ unstable in G. It is called <u>ordinary</u> if either

1) it is $\mathcal{L}^+$ and $\mathcal{L}^-$ stable

or

2) it is $\mathcal{L}^+$ stable and issuing from G as $t \to -\infty$.

From our remarks concerning critical points [cf. pp. 4 - 5] and the fact that for each limit-cycle $\bar{\gamma}$ of (A), $h(\bar{\gamma}) \neq 0$, we immediately obtain

LEMMA 6. The critical points, limit-cycles and separatrices of a system (A*) are singular trajectories.

Next we must prove

LEMMA 7. If $\bar{\gamma}$ is a limit-cycle solution of (A*) with $h(\bar{\gamma}) < 0$ [> 0], $\bar{\gamma}$ is $\mathcal{L}^+[\mathcal{L}^-]$ stable in G.

Let N_0 be a fixed point on $\bar{\gamma}$ and s the arc length from N_0 to another point N on $\bar{\gamma}$ measured positively in the direction of motion. We choose the units so that $\bar{\gamma}$ is of unit length. Then on $\bar{\gamma}$, $t = f(s)$, where f is monotone increasing and of period 1. Let $N = \Psi(s, N_0)$ be the motion of the representative point on $\bar{\gamma}$ and $M = \varphi(\sigma, M_0)$ the motion of the representative point on a nearby trajectory $\bar{\gamma}$. The position of M from $\bar{\gamma}$ [cf. [1]] is given by

$$\eta(s) = e^{hs}\,\Psi(s, N_0)$$

where η is the normal from M to a point N of $\bar{\gamma}$.

In view of the fact that we can choose M_0 such that the distance $\rho[M_0, N_0]$ is arbitrarily small, to prove the lemma it will suffice to show that the distance $\rho[M, N] \to 0$ as $s, \sigma \to \infty$.

Let us consider a point M in the vicinity of $\bar{\gamma}$ and let γ_δ be a curve through M parallel to $\bar{\gamma}$. We denote the arc length on γ_δ by σ_1. Let r(s) be the radius of curvature of $\bar{\gamma}$ and θ the angle of rotation. Then

$$ds = r d\theta\ ,$$
$$d\sigma^2 = d\sigma_1^2 + (d\eta)^2\ ,$$
$$d\sigma_1 = [r + O(\eta)]\, d\theta\ ,$$

here and elsewhere all the terms in $O(\eta)$ are uniform in θ. Whence we obtain

$$\begin{aligned} d\sigma^2 &= [r^2 + 2rO(\eta) + O(\eta)^2\ (d\theta)^2 + (d\eta)^2 \\ &= r^2[1 + 2O(\tfrac{\eta}{r}) + O(\tfrac{\eta}{r})^2]\, d\theta^2 + d\eta^2\ . \end{aligned}$$

Neglecting the terms of higher order

$$d\sigma = (r + rO(\tfrac{\eta}{r}))\, d\theta$$

and

$$|ds - d\sigma| = rO(\tfrac{\eta}{r})d\theta = O(\tfrac{\eta}{r})ds.$$

Integrating and passing to the limit as $s, \sigma \to \infty$ we obtain the desired result.

In view of the fact that the node, foci, and limit-cycles are either $\mathcal{L}^+$ or $\mathcal{L}^-$ stable, if we enclose these singular trajectories in critical regions and use the continuity theorem with respect to the initial conditions we obtain

LEMMA 8. The trajectories of type IV and V are ordinary.

To simplify our terminology we shall call the nodes, foci and limit-cycles sinks if they are $\mathcal{L}^+$ stable and sources if they are $\mathcal{L}^-$ stable. The notion of source is also extended so as to include the boundary L of G which is assumed to be a cycle without contact. Is several separatrices are entering G each arc of L lying between two separatrices will be considered as a distinct source.

LEMMA 9. The singular trajectories, that is, the sources, sinks, saddle points and separatrices, divide the region G into a finite number of components containing only ordinary trajectories.

Since there are only a finite number of singular trajectories in G and since through each non-critical point of G there is exactly one trajectory the singular trajectories divide G into a finite graph. Hence G contains only a finite number of components.

These components are of two classes: the interior class, that is, the class containing trajectories which are both $\mathcal{L}^+$ and $\mathcal{L}^-$ stable, and the exterior class, that is, the class containing trajectories which are issuing from L and are $\mathcal{L}^+$ stable, i.e. tending toward a sink.

We immediately observe that each component has at least one sink and one source on its boundary.

A sink and a source are connected if some of the trajectories issuing from the source tend toward the sink. Obviously two sources [sinks] cannot be connected.

THEOREM 7. Each component has exactly one sink and one source on its boundary.

In the proof of the theorem we shall assume that a component C does have more than one sink [source] on its boundary and derive a contradiction. To this end we shall show that if two sinks M_1, M_2 are on the boundary of the same component C, they must be connected.

We join M_1, M_2 by a rectifiable curve k lying entirely within C. For small values of the length parameter of k (as long as we are within the critical circle about M_1) each trajectory intersecting k must terminate at M_1.

Now the set of points on k through which there pass trajectories tending toward M_1 form an open set. For suppose M is a point of k through which there passes a trajectory tending toward M_1 and in every neighborhood of which there are trajectories which do not tend toward M_1. By the continuity theorem with respect to the initial conditions, all the trajectories passing through a sufficiently small neighborhood of M must enter the critical region containing M_1 and therefore terminate in M_1. This contradicts our assumption.

Hence the set of points on k connected with M_1 is open.

Let N be the first point on k which is not connected with M_1. A trajectory passes through N and since it cannot be a separatrix (k is entirely within C) it must terminate in two other sinks or sources M_3, M_4 (one may be M_2). But the set of points on k through which there pass trajectories connected with M_3 and M_4 is also an open set. Hence there are points on k before N through which there are trajectories not tending toward M_1. This contradicts our assumption that N was the first such point. Therefore all the trajectories passing through points of k tend toward M_1. However, if we follow k until we are within the critical region about M_2, the trajectories must also tend toward M_2. Therefore M_1 and M_2 are connected. This completes the proof of the theorem.

In view of Theorem 7, one can immediately deduce that in G a system (A*) has only components whose boundaries are made up of: (see Figure 3)

a) one source and one sink

b) one source, one sink, two saddle points and four separatrices.

c) one source, one sink, one saddle point and three separatrices (two of which are connected with the source [sink] the third with the sink [source].

The following lemmas show that if a system (A*) satisfies the condion (*) in G, then for δ sufficiently small, each system (B) also satisfies the conditions (*) in G.

In the statements and proofs of the lemmas we shall assume the critical regions containing the nodes and simple foci so small that the Jacobian $\Delta(x_1, x_2) > 0$ throughout the region. Similarly, we surround saddle points by simple closed curves such that $\Delta(x_1, x_2) < 0$ throughout the region.

In the statements of the lemmas we shall assume the critical regions (circles) containing the sources and sinks (saddle points) so small that

1) the distances of any point within the region (circle) from the source or sink (saddle point) is $< \delta < \dfrac{\varepsilon}{4S,\ e^{k\delta}}$ where S is the maximum of the lengths of the limit-cycles and k is the greater of the Lipschitz constants for P_1, P_2 in G,

and

2) throughout the regions (circles) containing critical points the Jacobian $\dfrac{D(P_1, P_2)}{D(x_1, x_2)}$ maintains the sign it possesses at the critical points.

Condition 1) implies that if R is a region containing a limit-cycle $\bar{\gamma}$ of (A*) and $\bar{\Gamma}$ is a limit-cycle of (B) also contained in R, then for δ sufficiently small, sign $h(\bar{\gamma})$ = sign $h(\bar{\Gamma})$.

LEMMA 9. If M_0 is a critical point of a system (A*) each system (B) has exactly one critical point and of the same type within a critical region R containing M_0.

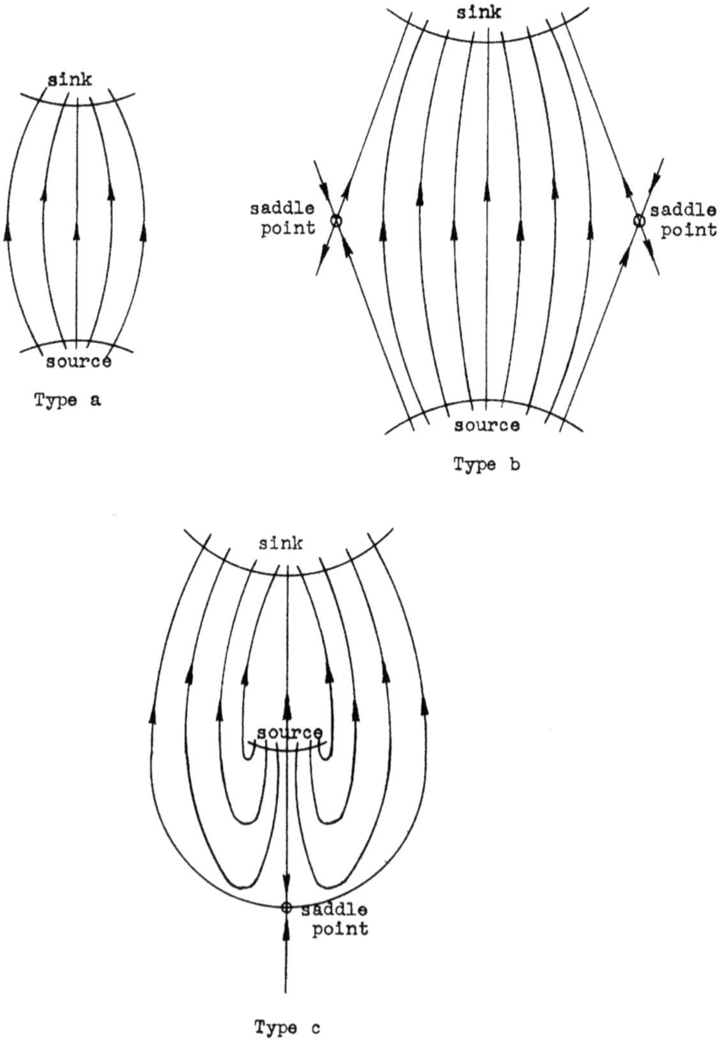

Figure 3
Types of Components

Suppose first that M_o is a node. Since, for δ small, for each (B) the index of the boundary of R [which is equal to the sum of the indices of the critical points contained in R] remains unchanged, R contains at least one critical point of each (B). Now $\Delta(x, y) > 0$ throughout R. Hence the critical points contained in R are either nodes or foci. But nodes and foci have index +1. Therefore R contains at most one critical point for each (B). We note too, that for δ small, the roots of the characteristic equation for each (B) are of the same type as those of (A*), i.e., real and of the same sign. Therefore the critical point of (B) in R is a node.

The same reasoning holds when M_o is a focus or saddle point.

LEMMA 10. If $\bar{\gamma}$ is a limit-cycle solution of (A*) and R a critical region containing $\bar{\gamma}$ but no other limit-cycle of (A*), then R contains exactly one limit-cycle of each (B).

To fix our ideas let us assume that $\bar{\gamma}$ is $\mathcal{L}^+$ stable, i.e. $h(\bar{\gamma}) < 0$. Since the trajectories of (A*) crossing the boundary of R are all directed to the interior of R those of each (B) will be similarly directed. Hence by the Bendixson-Poincaré theorem [6] R contains at least one limit-cycle solution of each (B). If R contained more than one limit-cycle, two say, for one (B), then on their facing sides the limit-cycles would be of opposite stability [ ]. This would imply that for this (B) at least one limit-cycle $\bar{\Gamma}$ would have the property that $h(\bar{\Gamma})$ would either vanish or change signs in R. But h is a continuous function of the perturbation. Hence for δ small, $h(\bar{\Gamma})$ must maintain the same sign as $h(\bar{\gamma})$. Thus our lemma is proved.

Let γ_1, γ_3 $[\gamma_2, \gamma_4]$ be separatrices of a system (A*) which issue from [tend toward] the sources [sinks] $\alpha_1, \alpha_3[\omega_2, \omega_4]$ respectively and tend toward [issue from] the saddle point S. Let c_1, c_3 denote the boundaries of the critical regions containing α_1, α_3 respectively and c the boundary of a circle of small radius enclosing S. By Lemmas 9 -10 there is exactly one source [sink] of (B) contained in each of these regions c_1, c_3 and one saddle point of (B) in c. We denote by Γ_1, Γ_3 $[\Gamma_2, \Gamma_4]$ the separatrices tending toward [issuing from] the saddle point $\mathcal{S}\varepsilon(B)$ which is contained in c.

LEMMA 11. The separatrices Γ_1, Γ_3 $[\Gamma_2, \Gamma_4]$ issue from [tend toward] the sources [sinks] of (B) which are contained in the critical region of α_1, α_3 $[\omega_2, \omega_4]$ respectively.

We denote the representative points on γ_1, γ_3 by $M^1 = \varphi(t - t_o^1, M_o^1)$, $M^3 = \varphi(t - t_o^3, M_o^3)$ respectively, where $M_o^1 = \varphi(0, M_o^1)$, $M_o^3 = \varphi(0, M_o^3)$ are the intersections of γ_1, γ_3 with c_1, c_3. Let $M_1^1 = \varphi(t_1 - t_o^1, M_o^1)$, $M_1^3 = \Psi(t_1 - t_o^3, M_o^3)$ be the intersection of γ_1, γ_3 with c. By the continuity theorem with respect to the initial conditions there are arcs η_1, η_3 of equal length on c_1, c_3 respectively which contain M_0^1, M_0^3 and such that all the trajectories of (A*) which intersect η_1, η_3 will after a time $\tau >$ max $[t_1^1, t_1^3]$ have entered c in the neighborhood of M_1^1, M_1^3. But since the only critical point of (A*) in c is a saddle point these trajectories will leave c after a certain time τ_2. Hence, on c there are two arcs μ_1, μ_3 containing M_1^1, M_1^3 respectively and such that all the trajectories of (A*) which intersect μ_1, μ_3 are entering c and two arcs μ_2, μ_4 not containing M_1^1, M_1^3 and such that all the trajectories of (A*) intersecting them are leaving c (cf. Figure 4).

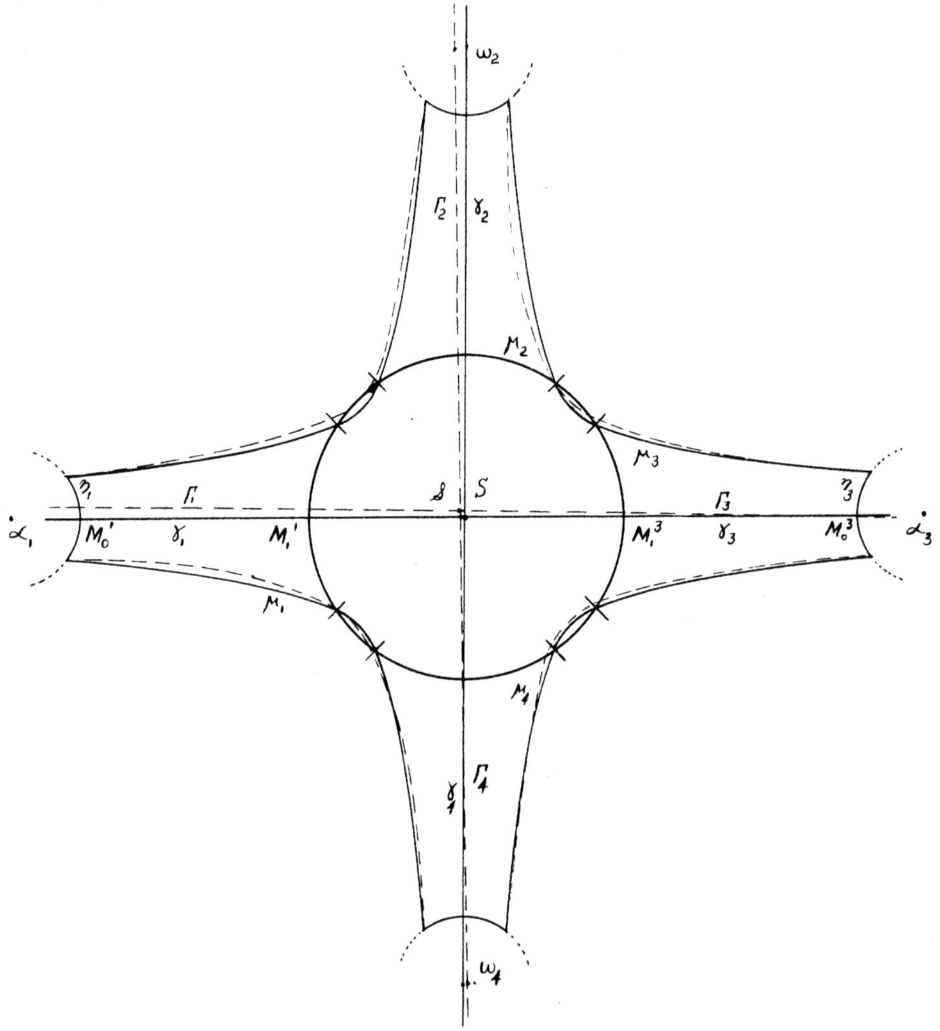

Figure 4

Now by Lemma 4, there exists a $\delta(\varepsilon)$ such that the trajectories of each (B) passing through the end points of η will remain arbitrarily close to the trajectories of (B) passing through these same points for τ, $t_0 \leq \tau \leq \tau_2$. Hence for each system (B) there are, on c, four arcs μ_i (i = 1, ..., 4) with properties analogous to μ_i (i = 1, ..., 4). Thus the trajectories of (B*) intersecting η_1, η_3 also form two open sets; viz, those that enter μ_1, μ_3 and leave by way of μ_2 and those that enter μ_1, μ_3 and leave by way of μ_4. Therefore η_1 and η_3 each contains a point through which a trajectory enters but does not leave c. Since the only critical point of (B) in c is the saddle point $\mathcal{S}$, these trajectories must be separatrices. This completes the proof of the lemma.

The lemma just proved can be restated in the following way:

> LEMMA 11'. Let c_1, c_2 be the critical circles containing the source and sink of a component C of type b. Let M_i (i = a, 2, 3, 4) be the intersection of the boundary of C with c_1, c_2. For each $\varepsilon_1 > 0$, there exists a $\delta(\varepsilon_1)$ such that the boundary of a component K of the system (B) intersects c_1, c_2 in $\mathfrak{m}_i$ (i = 1, 2, 3, 4) and the distance $\rho[M_i\, \mathfrak{m}_i] < \varepsilon_1$ (i = a, 2, 3, 4).

Before proceeding to the proof of our main theorem we make the following preliminary remark.

Each separatrix issuing from a saddle point on the boundary of components of type b, will be considered as the continuation of the separtrix tending toward the saddle point. The boundaries of the components of type c) have one separatrix γ issuing from [tending toward] the saddle point which two separatrices are tending toward [issuing from] the saddle point. Each side of γ will be considered as the continuation of the adjacent separatrix. In effect, this will reduce the consideration of components of type c) to those of type b). So that in the proof of Theorem 9 we need only consider components of types a) and b).

A system (A*) has i sources and j sinks in G, the two sides of the limit-cycles being counted as distinct sources or sinks. We denote the sources by α_m(M = 1, 2, ..., i) and the sinks by ω_n (n = 1, 2, ..., j). The components of (A*) having α_m, ω_n on their boundaries we denote by $C_\chi^{m,n}$(χ = 1, 2, ..., r(m, n)) where r(m, n) is the number of the components having both α_m, ω_n on their boundaries. Let c_m, c_n denote the boundary of the critical region containing α_m, ω_n respectively. The k separatrices which issue from c_m [tend toward c_n] divide c_m [c_n] into k arcs. We map each arc into the interval [0, 1] and obtain for each $C_\chi^{m,n}$ the one parameter family of trajectories.

$$\gamma^{m,n}_{\ell\ b} \quad \left\{ \begin{array}{ll} 0 \leq b \leq 1 & \text{for } k \geq 1 \\ 0 \leq b \leq 1\ ;\ \gamma^{m,n}_{\ell,0} = \gamma^{m,n}_{\ell,1} & \text{if } k = 0 \end{array} \right\}$$

Analogously, for each component $K_{\lambda}^{\mu,\nu} \epsilon (B)$

$$\begin{pmatrix} \mu = 1, 2, \ldots, i; & \nu = 1, 2, \ldots, j \\ \nu = 1, 2, \ldots, r(\mu,\nu) & \end{pmatrix}$$

we obtain the family

$$\left\{\Gamma_{\lambda,\beta}^{\mu,\nu}\right\} \begin{pmatrix} 0 \leq \beta \leq 1; & \text{if } k \geq 1 \\ 0 \leq \beta < 1; & \Gamma_{\lambda,0}^{\mu,\nu} = \Gamma_{\lambda,1}^{\mu,\nu} \quad \text{if } k = 0 \end{pmatrix}$$

We prove now

THEOREM 9. A system of type (A*) is structurally stable in G.

In the proof, first we shall show that there exists a trajectory preserving topological transformation of G onto itself and then we shall show that $\delta(\epsilon)$ can be chosen so small that $\rho[M, T(M)] < \epsilon$ $[M \epsilon G, T(M) \epsilon T(G)]$.

To avoid cumbersome notation we shall consider a typical component $C \epsilon (A^*)$ which is of type b. From Lemmas 9 - 12 there is, for each system (B) in the neighborhood of the boundary of C exactly one component $K \epsilon (B)$ of type b. We denote by c_α, c_ω the critical circles containing the sources and sinks, respectively, of (A*) and (B). First we assume the sources and sinks are not limit-cycles.

We denote the trajectories of C and K by

$$\{\gamma_b\} \ (0 \leq b \leq 1), \qquad \{\gamma_\beta\} \ (0 \leq \beta \leq 1) \ ,$$

respectively (Γ_0 lies in the neighborhood of γ_0) and the points of γ_b, Γ_β $(0 \leq b, \beta \leq 1)$ on c_α by M_b^0, m_β^0, respectively. Denote the arc lengths from M_0^0, M_1^0 to the saddle points on γ_0, γ_1 by s_0, s_1, respectively, and define the continuous function

$$s_b = (1 - b)s_0 + bs_1 \ (0 \leq b \leq 1).$$

This function gives a corresponding arc length from M_b^0 on each γ_b. Analogously, we define on the trajectories $\{\Gamma_\beta\}$ $(0 \leq \beta \leq 1)$ which belong to K the arc length function

$$\sigma_\beta = (1 - \beta)\sigma_0 + \beta\sigma_1 \ (0 \leq \beta \leq 1) \ .$$

Now map each trajectory of C and K into the open interval $(-1, +1)$; the points defined by s_b, σ_β $(0 \leq b, \beta \leq 1)$ corresponding to the mid-point of the interval. The representative points on γ_b, will then be given by the equations

$$M_b = \Phi(s + s_b, M_b^0) \ , \quad (-1 < s, \sigma < +1; \ 0 \leq b, \beta \leq 1)$$

$$m_\beta = \Psi(\sigma + \sigma_\beta, m_\beta^0)$$

respectively, where $M_b^0 = \Phi(0, M_b^0)$, $m_\beta^0 = \Psi(0, m_\beta^0)$ and

$$\lim_{s \to -1[+1]} \Phi(s + s_b, M_b^0) = \text{source [sin } \epsilon C$$

$$\lim_{\sigma \to -1[+1]} \Psi(\sigma + \sigma_b, \mathfrak{M}_\beta^0) = \text{source [sink]} \epsilon K .$$

The mapping

$$T: \quad \Phi(s + s_b, M_b^0) \rightarrow \Psi(\sigma + \sigma_\beta \mathfrak{M}_\beta^0)$$

for $s = \sigma$, $b = \beta$ $(-1 \leq s, \sigma \leq +1,\ 0 \leq b, \beta \leq 1)$ is a topological trajectory preserving mapping of C onto K. Clearly, if G contains no limit-cycles we extend this same mapping over all $C^{m,n}$ and $K^{\mu,\nu}$ and obtain a topological trajectory preserving mapping of G onto itself.

Now if G contains a limit-cycle $\bar{\gamma}$ of (A*), the critical region R containing $\bar{\gamma}$ contains exactly one limit-cycle $\bar{\Gamma}$ of (B). With obvious modifications of the mapping T just defined we can define a topological trajectory preserving mapping T_1 of $G - R$ onto itself.

We denote the trajectories of (A*) which intersect $\gamma_{+\delta}, \gamma_{-\delta}$ by

$$\{\gamma_b\}\ ,\ \{\gamma_b'\} \quad (0 \leq b < 1;\ \gamma_0 = \gamma_1,\ \gamma_0' = \gamma_1')\ ,$$

respectively, and the trajectories of (B) by

$$\{\Gamma b\}\ ,\ \{\Gamma_\beta'\} \quad (0 \leq \beta < 1; \Gamma_0 = \Gamma_1, \Gamma_0' = \Gamma_1')$$

where $T_1(\gamma_0) = \Gamma_0$. We denote the normals to $\bar{\gamma}$ by

$$\{\eta_a\} \quad (0 \leq a < 1;\ \eta_0 = \eta_1)\ ,$$

a increasing in the sense of positive motion.

It is well known [3] that the intersections of each trajectory of (A*) with each η_a are sequences of points tending toward the intersection $\eta_a - \bar{\gamma}$ as a limit. Let the integer $i = \pm 1, \pm 2, \pm 3, \ldots,$ denote the subsequent intersections of each γ_b, γ_b' with each η_a (the integers > 0 denoting the intersections with γ_b and those < 0 the intersections with γ_b'). The points on γ_b within R may now be given by the coordinates (a, b, i) $(0 \leq a \leq 1;\ i = 1, 2, 3, \ldots)$.

We note that for δ small in the region R the velocity vectors of (B) have approximately the same direction as those of (A*). Hence the trajectories of (B) also intersect the η_a's with a definite angle different from zero. Let $j = \pm 1, \pm 2, \pm 3, \ldots,$ be the analogue of i for the trajectories of (B).

Now under T_1', $(a, b, +1) \rightarrow (\varphi^+(a), b = \beta, +1)$ and $(a, b, -1) \rightarrow (\varphi^-(a), b = \beta, -1)$. Under the assumption that T_1 is a trajectory preserving ε-mapping of $G - R$ onto intself, φ^+, φ^- are both monotone increasing.

We define

$$T_2: (a, b, i) \rightarrow (\alpha, \beta, j)(0 \leq b, \beta, a, \alpha \leq 1;\ i.j = \pm 1, \pm 2, \pm 3, \ldots)$$

whenever $b = \beta$, $i = j$ and

$$\alpha = \frac{1}{2}(1 + \frac{1}{i})\,\varphi^{+}(a) + \frac{1}{2}(1 - \frac{1}{i})\,\varphi^{-}(a) .$$

This mapping is clearly continuous and trajectory preserving. It is also one-one. For suppose, if possible, that two values of $\underline{a}$ viz., a_1, a_2 $(a_2 > a_1)$ yield the same value α. Then

$$(1 + \frac{1}{i})\,[\varphi^{+}(a_1) - \varphi^{+}(a_2)] = (1 - \frac{1}{i})\,[\,\varphi^{-}(a_2) - \varphi^{-}(a_1)].$$

Since φ is monotone increasing, the left side of this equality is zero for $i = -1$ and < 0 otherwise. While the right side is zero for $i = +1$ and > 0 otherwise. Thus T_2 is one-one and the mapping $T = T_1 + T_2$ is a topological mapping of G onto itself which preserves trajectories. We add the remark that if the distance $\rho[M, T_1(M)] < \varepsilon/4$, then T_2 is a trajectory preserving ε-mapping of R onto itself.

To complete the proof we have only to show that for each $\varepsilon > 0$ we can find a $\delta(\varepsilon)$ such that the distance $\rho[M, T(M)] < \varepsilon$ $[M \in G, T(M) \in TG]$.

First we construct the critical regions such that each point in each region is a distance $< \varepsilon/4$, say, from the source or sink contained in the region. By Lemma 11 we choose $S = S_1$ so that the distance

$$\rho[M_0, \mathfrak{M}_0^0],\quad \rho[M_1, \mathfrak{M}_1^0] \text{ are less than } \varepsilon/4e^{k\bar{s}}$$

where $\bar{s}$ is the maximum of the arc lengths along the trajectories between the boundaries critical regions containing the source and sink and K the maximum of Lipschitz constants for P_1, P_2 in G. Then the distance $\rho[M_b^0, \mathfrak{M}_\beta^0] < \varepsilon/2e^{k\bar{s}}$ for $b = \beta(0 \leq b, \beta \leq 1)$. Now choose $\delta < \min[\delta_1, \varepsilon/2e^{k\bar{s}}]$. Then by Lemma 4, $\rho[M, T(M)] < \varepsilon$ for all points outside the critical regions. But the critical regions were chosen so small that the distance of any point within the region is less $< \varepsilon/4$ from the source or sink. Therefore $\rho[M, T(M)] < \varepsilon$ also within the critical regions. This completes the proof of our theorem.

The treatment of structural stability with respect to parametric variations [1] is entirely unsatisfactory. For this reason we include here the following remark which may serve to clarify the situation somewhat. A more complete discussion of this problem will appear elsewhere.

We consider systems of type

(A) $$x = P_i(x_1, x_2; \lambda)\quad (i = 1,2)$$

where P_1, P_2 are analytic with respect to the variables x_1, x_2 and the parameter λ within and on the boundary of a region G. We assume L, the boundary of G, to be a cycle without contact.

DEFINITION: The system (A) is structurally stable in G with respect to $\lambda = \lambda_0$ if for each $\varepsilon > 0$ there exists a $\delta(\varepsilon) > 0$ such that for all $\lambda (\lambda_0 - \delta \leq \lambda \leq \lambda_0 + \delta)$ there exists a topological transformation T of G onto itself with the properties

1) Trajectories of

(A) $$\dot{x}_i = P_i(x_1 \ x_2; \lambda_0) \quad (i = 1, 2)$$

map onto trajectories of

(B) $$\dot{x}_i = P_i(x_1 \ x_2; \lambda) \quad (i = 1, 2)$$

and

2) $[M, T(M)] < \varepsilon \quad (M \in G,\ T(M) \in T(G)).$

All the theorems and lemmas of Part II of this paper hold for systems of type (A*). Hence the conditions (*) are certainly sufficient to guarantee the structural stability of systems of type (A).

BIBLIOGRAPHY

[1] ANDRONOV A., CHAIKIN, C., "Theory of Oscillations," - English Language Edition, Princeton University Press, 1949.

[2] ANDRONOV, A., PONTRJAGIN, L., "Comptes Rendus," (Doklady), Acc. Sc. U. S. S. R., Vol. 14, 1937.

[3] BENDIXSON, I., "Acta Mathematica," Vol. 24, 1901.

[4] DULAC, H., "Bulletin de la Scoiete Math. de France," Vol. 51, 1923.

[5] HUREWICZ, W., "Ordinary Diff. Equations in the Real Domain," Brown University Notes, Autumn, 1943.

[6] LEFSCHETZ, S., "Lectures on Diff. Equations," Princeton University Press, 1948.

NOTE: This paper was prepared under contract with the Office of Naval Research, and was equally sponsored by the Office of Air Research.

NOTES ON DIFFERENTIAL EQUATIONS

By Solomon Lefschetz

The two notes which together make up the present paper deal with material given in my Princeton course on Differential Equations of the fall semester of 1949. While the notes are independent they are not unrelated; and the topics discussed: <u>critical points</u> and <u>the equation of van der Pol</u> are of continuing interest in differential equations.

In both notes we deviate from standard terminology by saying "critical point" and "path" instead of "singular point" and "trajectory".

The references [ ] are to the bibliography at the end of the paper.

I. ON CRITICAL POINTS

1. Our present objective is to present a topological analysis of the critical points of a real system

$$\frac{dx}{dt} = P(x, y), \quad \frac{dy}{dt} = Q(x, y) \tag{1.1}$$

where P and Q are analytical wherever considered. The critical points are the points where $P = Q = 0$. We will only consider isolated critical points. Thus if 0 is a critical point under consideration there is a circular region Ω of center 0 in which P and Q are both holomorphic and vanish simultaneously solely at the point 0.

Critical points of systems (1.1) and under more general assumptions have been dealt with repeatedly in the literature. Indeed a very complete treatment is contained in the recent book by Stepanoff-Niemitzki ([3], Ch. II, §§ 3, 4), (also in the earlier 1947 edition). Our approach is, however, exceptionally simple and direct. It is true that we only consider analytical systems but it is also clear that much weaker assumptions could be made at little cost.

2. We will first discuss two preliminary lemmas.

(2.1) LEMMA. Let $f(x, y)$ be a real function holomorphic in a circular region Ω of radius ρ and center at the origin. If $f(0, 0) = 0$ and ρ is sufficiently small, then the real curve $f(x, y) = 0$ has in Ω at most a finite number of real branches which are of one of the two forms

(2.1a) $x = 0$

(2.1b) $y = a_k x^{k/n} + a_{k+1} x^{\frac{k+1}{n}} + \dots,$

where $n, k > 0$ and the coefficients are all real. Moreover for ρ small enough the branches interact one another in Ω solely at the origin.

According to the Weierstrass preparation theorem we have in a suitable region:

$$f(x, y) = E(x, y) \cdot g(x, y) \cdot x^q, \quad q > 0$$

$$g(x, y) = y^m + a_1(x) y^{m-1} + \dots + a_m(x)$$

where all functions are holomorphic in Ω, $E \neq 0$ in Ω, and $a_i(0) = 0$. Thus the solutions of $f = 0$ in Ω are $x = 0$ if $q > 0$ and the solutions of $g = 0$. By Puiseux's theorem [5, p 421] the latter have the form (2.1b) and (2.1b) is only real when all the a_k are real. The intersections of two branches, say (2.1b) and another with k', n', a'_k in place of k, n, a_k, correspond to the solutions in Ω of a relation

$$x^{r/nn'}(b_0 + b_1 x^{1/nn'} + \dots) = 0$$

where the b_j are not all zero. By a well known property of analytic functions the solutions $\neq 0$ are bounded away from zero, and hence not in Ω for ρ sufficiently small. This completes the proof of the lemma.

We will now consider a region Ω in which P and Q of (1.1) are holomorphic and have at most a finite number of critical points. The parts of paths of (1.1) and other arcs considered below are always assumed to lie in Ω.

Let A, B be the end-points of an arc λ. It will be convenient to use the designations

$$(AB), \quad [AB), \quad (AB], \quad [AB]$$

to denote respectively the open arc λ; λ closed at A, open at B; λ open at A, closed at B; λ closed at both ends.

We recall now the following definition due to Poincaré. Relative to the system (1.1) an arc λ (open or closed at one or the other end) is said to be <u>without contact</u> wherever λ contains no critical points and the (unique) path γ through any point M of λ is never tangent to λ at M.

(2.2) LEMMA. Let $\lambda = (AA')$ and $\mu = (BB')$ be two open arcs without contact disposed as in Figure 1 so that A and B are on the same path γ_0. Given any point $M \in \lambda$ let the path $\gamma(M)$ through M when followed forward first meet μ in a point N and so that from M to N the path does not cross λ. Then $M \to N$ defines a topological mapping T of $[AA')$ on a subarc $\lambda' = [BA'')$. Moreover if $B' \neq A'$, hence $A'' \neq A'$, then the assignment $A'' = TA'$ makes of T a topological mapping $[AA'] \to [BA'']$. The same properties hold if the paths are followed backward instead of forward.

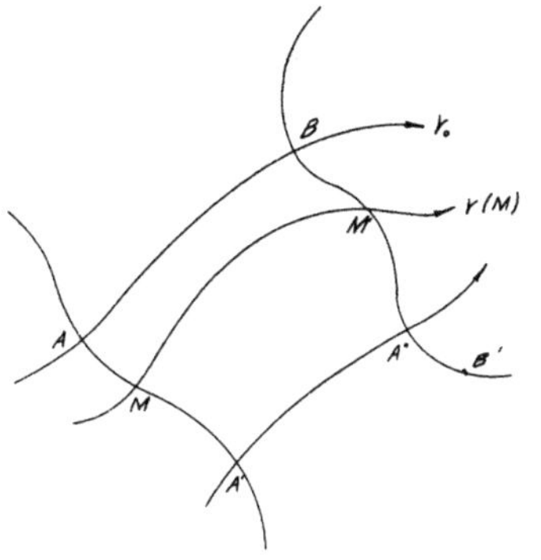

Figure 1

It is clear that T straight or extended is one-one. Its bicontinuity is a well known consequence of the existence theorem for systems such as (1.1).

3. We are now ready to discuss the nature of the critical points of the system (1.1). We suppose then that the origin 0 is an isolated critical point and Ω a circular region with the origin as center containing no other critical point than 0 and in which P and Q are analytic. It is understood once for all that we are operating throughout within Ω.

Let r, θ denote the polar coordinates of the point (x, y). Along the path through the point we have from (1.1)

(3.1) $$r\frac{dr}{dt} = xP + yQ = R(x, y).$$

Let us suppose first that the origin is not an isolated point of the curve $R = 0$. Then there is at least one branch of the curve through 0 and one may assume Ω such that the branches intersect only at 0 in Ω. It follows that a small neighborhood of 0 is divided into sectors in which dr/dt has a fixed sign. Let us take one of the sectors and close it, to form a triangular region OAB (Figure 2) where AB is an arc of a circle of radius ρ and center 0 contained in Ω.

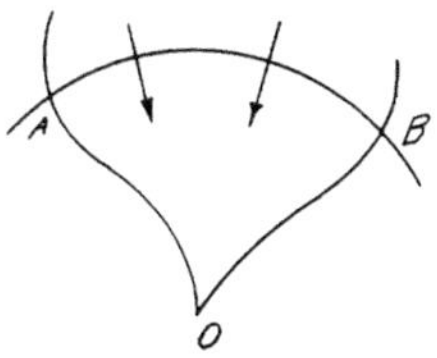

Figure 2

It follows from our construction that (AB) is an arc without contact. Let us show that:

(3.1) <u>If ρ is sufficiently small, then (OA] and (OB] are without contact.</u>

It is sufficient to consider, say (OA]. Let the coordinates be so chosen that none of the branches of the curve $R(x, y) = 0$ is tangent to one of the axes. Now the branch (2.1b) has for tangent

$$\begin{array}{ll} x = 0 & \text{if } k < n \\ y = 0 & \text{if } k > n \\ y = a_k x & \text{if } k = n. \end{array}$$

We are then here in the third case. Hence our branch has a representation

$$y = mx + ax^n + \dots, \quad m \neq 0.$$

At any point (x, y) of $(OA]$ the tangent to the branch has for slope

$$\mu = m + nax^{n-1} + \dots$$

while the path through the point has the same slope μ' as the circle through the point. Thus

$$\mu' = -\frac{x}{y} = \frac{-1}{m+ax^{n-1}+\dots} .$$

Hence for ρ small enough

$$\mu\mu' = -1 + \varepsilon, \text{ where } \varepsilon \to 0$$

with ρ. Thus for ρ sufficiently small certainly $\mu' \neq \mu$, and this implies (3.1).

To sum up then along each side of the curvilinear triangle OAB the crossing of the paths, whether inward or outward is fixed.

4. Let us suppose first that the side AB is crossed inward by the paths, i.e. that ρ decreases along all paths in the sector under consideration. There are then various possibilities depending upon the modes of crossing of the other two sides. Each case must be examined separately.

Case I. The sides OA and OB are both crossed outward. A path γ through a point M of (OA) when followed backward necessarily goes farther from the vertex O throughout the triangle and must therefore cross (AB) in a point N. By lemma (2.1) $N \to M$ defines a topological mapping of [AO] on a subarc [AC] of [AB] and the path δ_1 through C tends to O (Figure 3). Similarly for the paths through the points of (OB) and a certain path δ_2. The paths crossing (AC) behave like γ, those crossing [CD] behave like δ, and those crossing (DB) behave like ε. For evident reasons the paths crossing [CD] form a system referred to as a fan.

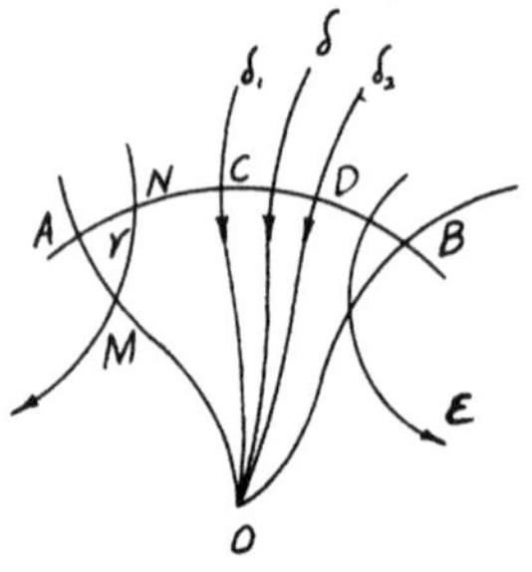

Figure 3

As a limiting case we might have C = D, hence $\delta_1 = \delta_2$ so that the fan consists of a single arc. This is shown in Figure 4.

Case II. The sides OA and OB are both crossed inward. Thus the whole periphery of the triangle, O excepted is crossed inward. Since the triangle contains no limit cycle or critical point, by the Poincaré-Bendixson theory, all the paths considered must tend to O, giving the disposition of Figure 5.

Case III. The two sides, OA, OB are crossed in opposite manner. Assume that OA is crossed inward and OB outward. The path δ through A may then either remain in the triangle, and then it can only tend to O, or else leave it and hence cross (OB). This gives rise to two subcases.

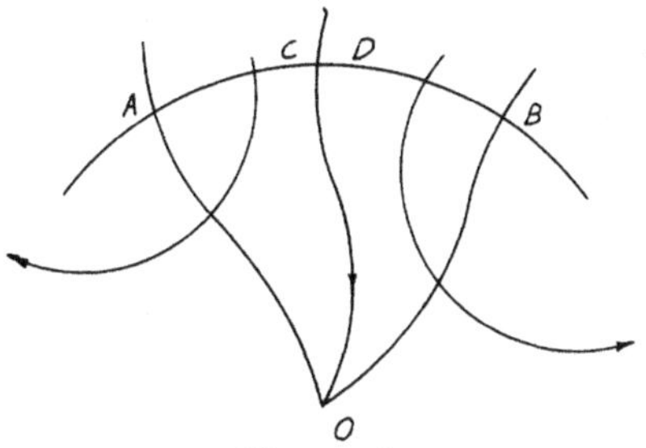

Figure 4

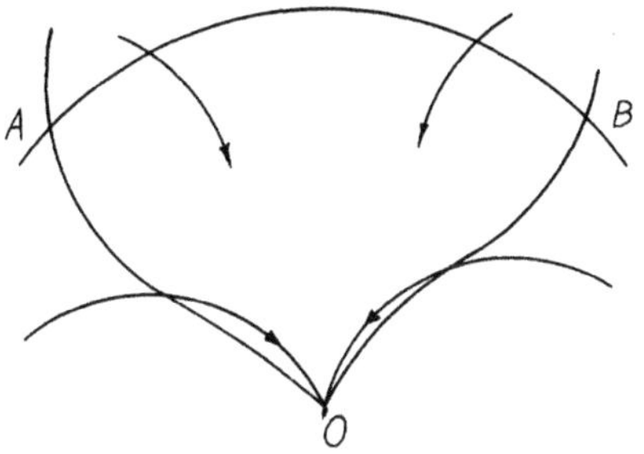

Figure 5

(a) *The path δ through A tends to the vertex O in the triangle.* As under Case I, the paths through the points of (OB) give rise to a topological mapping of [OB] onto a subarc [CB] of [AB] (Figure 6). The paths through [AC] form a fan. The paths γ, δ, ε exhibit the three types of paths in relation to the triangle.

(b) *The path δ through A crosses (OB).* We have then the situation of Figure 7, with only two typical paths: γ and δ.

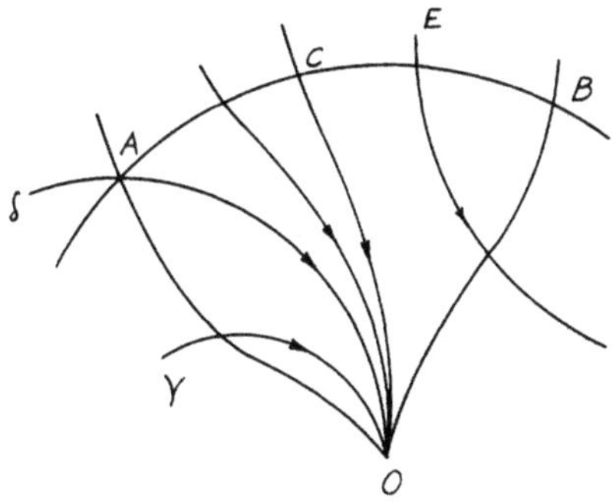

Figure 6

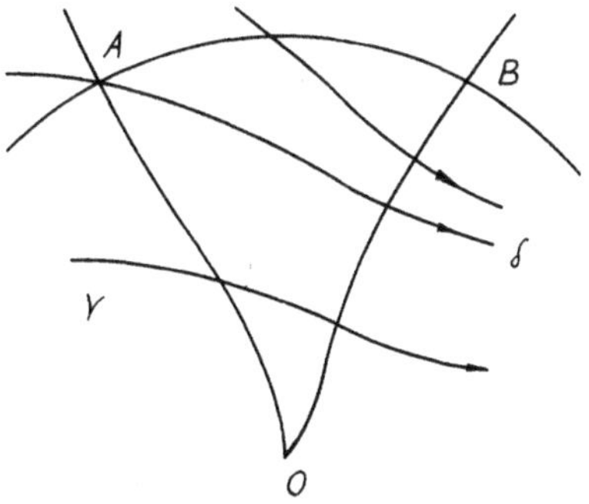

Figure 7

If the crossings at OA and OB are reversed there will result cases III_1a, ..., derived from IIIa, ..., in the obvious way.

5. It has been assumed so far that (AB) is crossed inward by the paths. If the crossing is outward the change of time variable $t \to -t$ will preserve the paths but make them cross (AB) inward. The resulting configurations are the same as before but with all the arrows reversed. We shall refer to them as I', ...

Suppose now that the curve $R = xP + yQ = 0$ has no real branches through the origin. Thus in one vicinity of the origin dr/dt has a constant sign. Let us suppose first $dr/dt < 0$ so that in the vicinity of 0, r is decreased along any path γ. Consequently along γ the coordinate r tends to a limit r_0 and $r = r_0$ is a limit-cycle of diameter $\leqq \rho$. Since we are dealing with an analytical system, for ρ small enough there is no limit-cycle in the circle C of radius ρ. Thus the only possibility is $r_0 = 0$, i.e. γ tends to 0.

Suppose that there is a ray L through the origin which is not a path. Then as regards the open sector of angle 2π, bounded by L the situation is that of case III. We have now the two possibilities corresponding to IIIa and IIIb which we consider in turn.

Case IV. One of the paths δ crossing L reaches the origin in the open sector. Then by our earlier discussion it will be seen that all the paths near enough to 0 behave like δ. Thus the point 0 is a stable node. (Figure 8).

Case V. Every path δ crossing L, say at a point M_1, crosses it again at a point M_2. For another path δ' as in Figure 9, the crossings will be M'_1, M'_2 and by lemma (2.1) $M'_1 \to M'_2$ defines a topological mapping $[OM_1] \to [OM_2]$, which shrinks the interval OM_1. It follows that δ, and hence every path sufficiently near the origin 0 is a spiral tending to 0. Thus 0 is a stable focus. Exceptionally $M' = M'_2$ throughout OM_1 and 0 is a center.

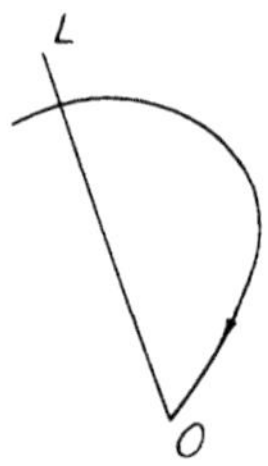

Figure 8

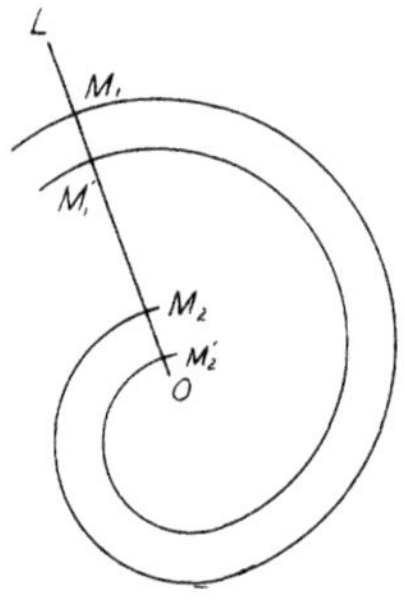

Figure 9

We have in addition

Case VI. $dr/dt = 0$, hence all paths are circumferences of center 0. The origin 0 is then again a center.

If every ray issued from 0 is a path, we have actually Case IV and 0 is a node.

Finally if $dr/dt > 0$ near 0 we have reversal of arrows and Case IV': unstable node, and Case V': unstable focus.

6. To obtain the full configuration of the paths around the critical point 0 (the local phase-portrait around the point) we have to combine adjacent sectors of the various types described in the various admissible ways.

Let us observe at the outset that if all the sectors around 0 are of types IIIb (Figure 7) or III_1b limit (rotation reversed) or the same with III' instead of III, then all the paths near 0 are spirals and we merely have a stable or unstable focus. As this case will arise otherwise anyhow we may leave out of consideration at present. On the other hand, a succession of sectors such as IIIb or III_1b not surrounding fully the point 0, has no topological effect on the configuration of the paths. Thus the situation of Figure 7 and related types need not be considered here. Upon matching the other types in the various admissible ways, and adding to them the focus and center we obtain the following list:

I. *Fan*. This may be *attractive* or *repulsive* accordingly as as the paths all tend to the critical point or all away from it (Figure 10). As a limiting case we have the *nodes*.

II. *Hyperbolic type*. (Figure 11).

III. *Nested ovals*. (Figure 12). This is the only truly new type.

IV! *Focus*.

V. *Center*.

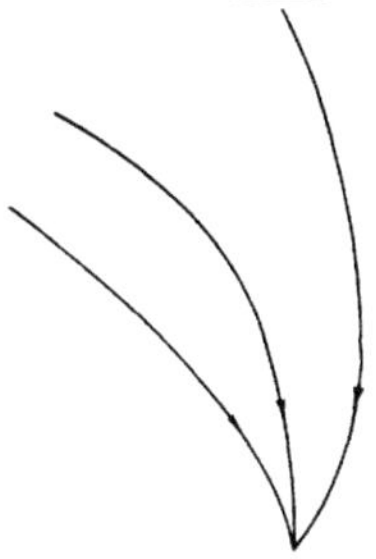

Figure 10

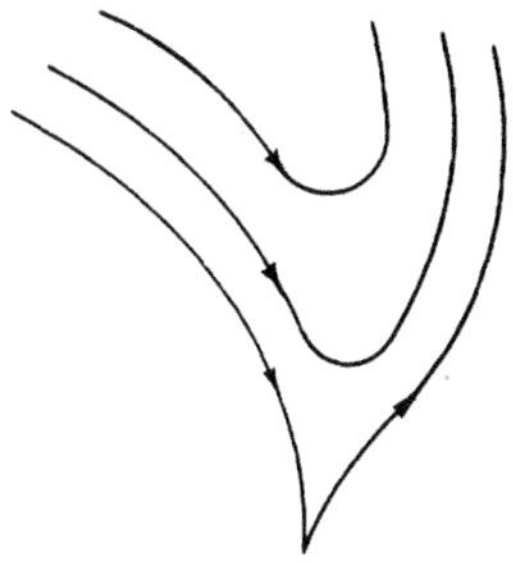

Figure 11

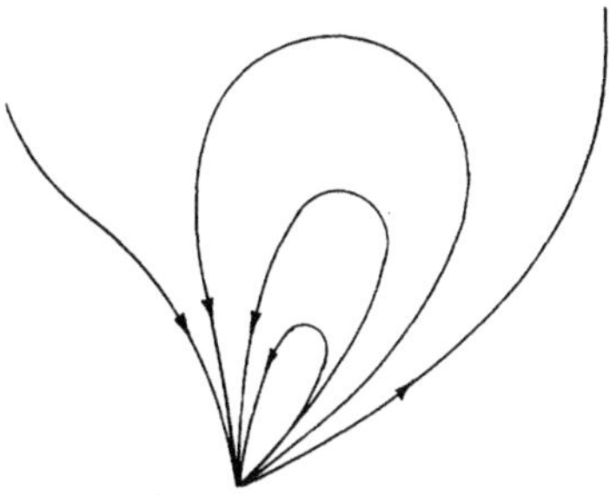

Figure 12

Existence of the various types. All the types except III are known to occur as critical points of linear equations. Regarding III consider a system of concentric circles as paths in the projective plane π obtained from the phase plane in the usual way. Let L be a line through the center O and let π be subjected to a quadratic transformation [5, p 431] relative to a real triangle ABC such that O is on the side BC but is not one of the vertices B, C. The circles will then be transformed into a system of paths with two sets of nested ovals relative to the critical point A.

II. ON THE EQUATION OF VAN DER POL

7. The equation of van der Pol

7.1 $$\ddot{x} + \lambda(x^2 - 1)\dot{x} + x = 0, \lambda > 0,$$

(dots denote time derivatives) is by now classical. By introducting the new variable

(7.2) $$y = \dot{x} + \lambda(\frac{x^3}{3} - x),$$

one may replace (7.1), in well known manner, by the equivalent system

(7.3) $$\dot{x} = y - \lambda(\frac{x^3}{3} - x),$$
$$\dot{y} = -x.$$

About all that is known regarding the "phase-portrait" (full picture of the paths) of this system is that there is a unique stable limit-cycle Γ (Liénard) and that the origin is the only critical point. This point is always unstable and is a node for $\lambda \geqq 2$ and a focus for $\lambda < 2$. From general theory one knows then that all the paths interior to Γ, and all the paths sufficiently near Γ and exterior to it spiral towards Γ (here in clockwise direction). Curiously enough and in spite of the importance of the van der Pol equation, the full phase-portrait of the system (7.3) has never been analyzed, or at least it is not found in the literature. It will be seen that, as one might well expect, a discussion of the singular points "at infinity", after the manner of Poincaré, provides most of one's required information. For complete details regarding this technique and more ample bibliographical data see [4].

8. Considerable information regarding the phase-portrait may be gleaned directly from the system (7.3). At the crossings with the y axis the tangents to the paths are horizontal, and at the crossings with the curve,

$$\Delta: \quad y = \lambda\left(\frac{x^3}{3} - x\right)$$

they are vertical. We see also from (7.3) that with increasing time:

x increases above Δ,
x decreases below Δ,
y increases at the left of the y axis,
y decreases at the right of the y axis.

Hence the behavior of the paths at the crossings with the y axis and the curve Δ is as indicated in Figure 13. Remember also that, according to Liénard, the limit cycle Γ is related to the curve Δ after the manner of Figure 13, and is stable.

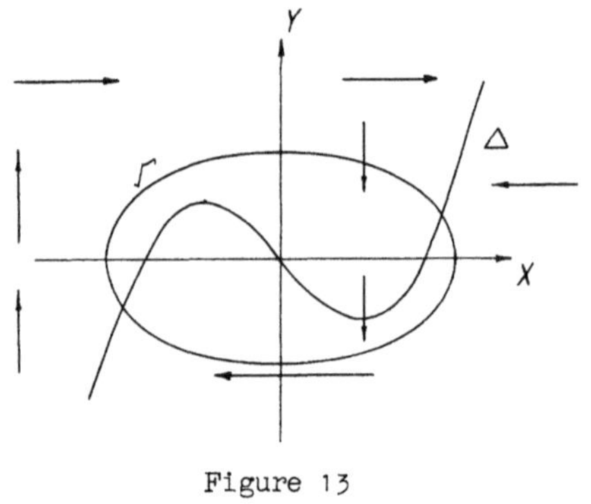

Figure 13

9. We must now study the behavior of the paths at infinity, in the phase plane. For that purpose we must have recourse to the Poincaré procedure. Let us first recall its main geometric features.

Let x, y, z be coordinates for an Euclidean 3-space E^3 and let the phase plane $E^2(xy)$ of the system (7.3) be identified with the plane $z = 1$ of E^3. If one completes E^2 to a projective plane π then the points of π are in one-one correspondence with the lines of E^3 through the origin O. If M is any point of π then OM intersects the sphere

S: $x^2 + y^2 + z^2 = 1$, in two points M', M", where M' is in the upper hemisphere S': $z \geq 0$, and M" in the other S". What one does actually in the Poincaré scheme is to study the paths described by the points M', M" on the whole sphere S. The phase-portrait in one of the open hemispheres is then the topological image of the desired phase-portrait in E^3.

Now on the sphere S the paths are those of the differential equation

$$\begin{vmatrix} dx, & dy, & dz \\ x, & y, & z \\ yz^2 - \lambda(\frac{x^3}{3} - xz^2), & -xz^2, & 0 \end{vmatrix} = 0,$$

which upon expanding the determinant yields

$$xz^3 dx + z\{yz^2 - \lambda(\frac{x^3}{3} - xz^2)\}\ dy$$

(9.1)

$$= x^2z^2 + y\{yz^2 - \lambda(\frac{x^3}{3} - xz^2)\}\ dz.$$

While z does not factor out nevertheless (9.1) is satisfied by $z = dz = 0$. Hence the arcs of the circle $z = 0$ between critical points are paths of (9.1).

Let us now look for the critical points on the line $z = 0$. Consider first the points for which $x \neq 0$. We may thus take $x = 1$, $dx = 0$ and we have to find if the points $y = z = 0$ (points at infinity on the x axis) are critical for

$$z\{yz^2 + \lambda(z^2 - \tfrac{1}{3})\}\ dy = y^2z^2 + y\{yz^2 + \lambda(z^2 - \tfrac{1}{3})\}\ dz.$$

The "critical point" behavior is governed by the linear approximation

(9.2) $$z\ dy = y\ dz.$$

As the general solution of (9.2) is $y = Cz$, the origin is the simplest type of node for (9.2): paths tending to the node in every direction. Hence it is a node of the same type for (9.1) and similarly for the two associated points on the sphere. In the xy plane the curves tending to infinity in the direction of the x axis will have horizontal asymptotes. Upon referring to the mode of increasing or decreasing of y it will be seen that each curve tends to its asymptote from above in quadrants II, III and from below in quadrants I and IV (Figure 14).

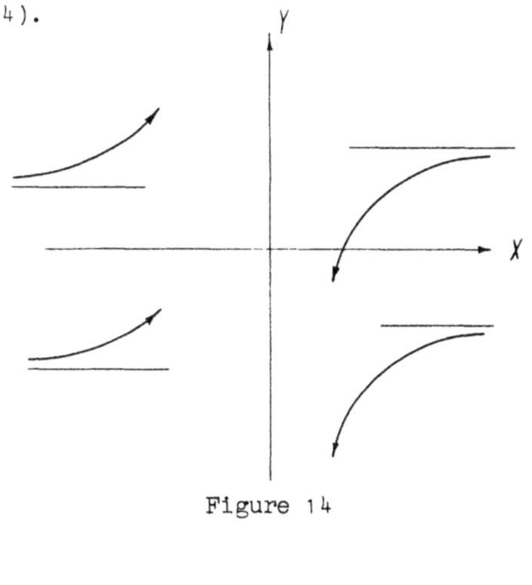

Figure 14

Assuming similarly $y = 1$, $dy = 0$ we look for the critical points of

$$xz^3dx = x^2z^2 + z^2 - \lambda\left(\frac{x^3}{3} - xz^2\right) \quad dz \tag{9.3}$$

and find only the points $x = z = 0$. Thus the only critical points of (9.3) are at infinity on the axes.

The behavior at the origin is the same as for

$$\frac{dx}{d\tau} = x^2z^2 + z^2 - \lambda\left(\frac{x^3}{3} - xz^2\right), \quad \frac{dz}{d\tau} = xz^3 . \tag{9.4}$$

Since there are no first degree terms the origin is a critical point of higher order. One could apply the analysis of I, but it is actually simpler here to proceed directly.

10. Let us examine the behavior of the paths outside of the limit cycle. Take a path γ beginning to the left of the y axis (Figure 15).

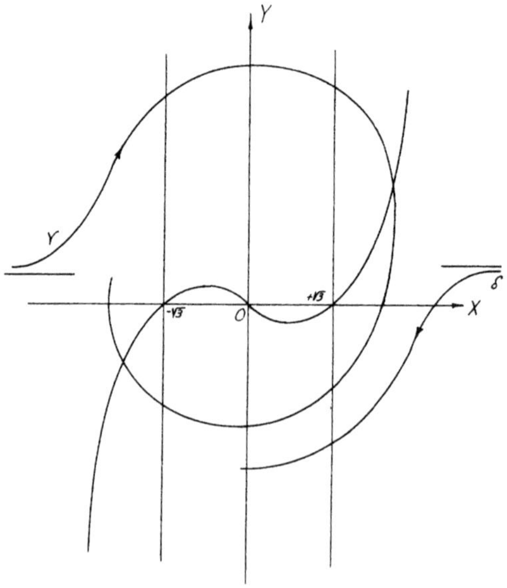

Figure 15

It has a horizontal asymptote and tends to it from above. We have on any path

$$r\frac{dr}{dt} = - \lambda x^2\left(\frac{x^2}{3} - 1\right) .$$

Hence for $x^2 > 3$, $\frac{dr}{dt}$ is negative and so γ rises but comes nearer to the origin till it reaches the line $x = - 3$. On the other hand

$$y' = \frac{dy}{dx} = - \frac{x}{y - \left(\frac{x^3}{3} - x\right)}$$

$$y'' = a \frac{y - \lambda(\frac{x^2}{3} - x) - x \quad \frac{-x}{y - \lambda(\frac{x^3}{3} - x)} - \lambda(x^2 - 1)}{(y - \lambda(\frac{x^3}{3} - x))^2}$$

Hence for y large and positive and $x^2 < 3$, approximately $y'' = -\frac{1}{y} < 0$. Hence within those limits the concavity of the curve points downward. The curve rises up to the y axis. After that it descends steadily to the x axis. Then it must maintain itself steadily above a curve such as δ . The latter by the same considerations as for γ behaves as in Figure 15 and so γ returns to the left x axis. All the paths starting to the left x axis between γ and the limit-cycle behave necessarily like γ , since they must proceed without intersecting γ. In particular γ itself continued can only wind spirally on the limit-cycle.

Consider now the situation on the sphere (Figure 16). The path δ' cuts OA in a point H distinct from A. The path γ' cuts Δ' and OA at points P and Q. Since no path is tangent to FB nor to OA, $P \rightarrow Q$ determines a topological mapping of [FB] onto a segment $[GG_1]$ where G_1 is between G and H. The path ε through G_1 tends to B tangentially to the circle C and it is a separatrix issued from B in the first quadrant.

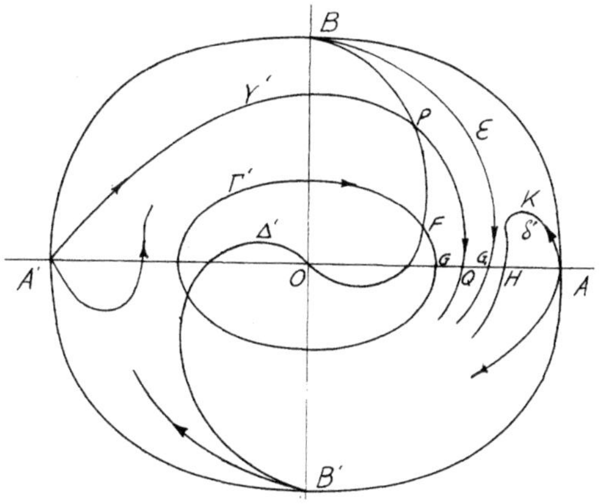

Figure 16

We shall show by an argument due to John McCarthy that ε is unique. Suppose in fact that there exists a second ε , (Figure 17) analogous to ε_1 . If $x(z)$ and $x_1(z)$ are the solutions of (9.3) corresponding to ε and ε_1 then

$$\frac{dx}{dz} = \frac{x}{z} + \frac{1}{xz} - \lambda(\frac{x^2}{3z^3} - \frac{1}{z})$$

and similarly for ε_1. Hence

$$\frac{d(x - x_1)}{dz} = \frac{x - x_1}{z} - \frac{(x - x_1)}{x\, x_1\, z} - \frac{\lambda(x^2 - x_1^2)}{3z^3} .$$

Hence setting $|x - x_1| = u$ we find

$$\text{(10.1)} \qquad \frac{d \log u}{dz} = \frac{1}{z} - \frac{1}{x\, x_1\, z} - \frac{\lambda(x + x_1)}{3z^3} .$$

Now both x and x_1 are of order < 1 in z. Hence in (10.1) the last two terms dominate the first and so

$$\frac{d \log u}{dz} < 0$$

for z small. Hence u increases when z decreases. Since this contradicts the disposition of Figure 17, ε_1 cannot exist and ε is unique.

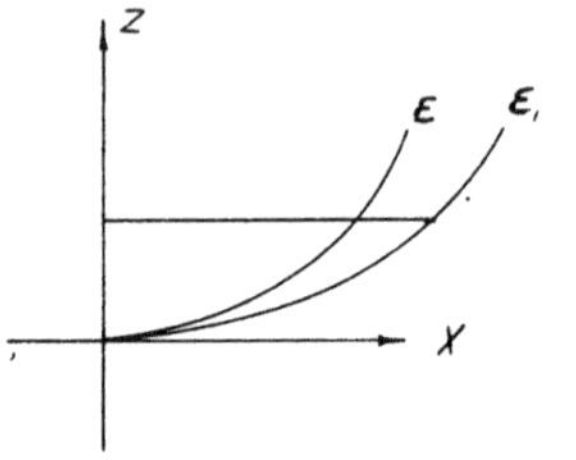

Figure 17

It follows from what precedes that to the right of BG_1 the paths can only emanate from A and so they all behave like δ'. The disposition of the paths in quadrants IV and III is then obtained by symmetry and they are disposed outside the limit-cycle, as in Figure 16. Within the limit-cycle the paths all spiral toward the limit-cycle. As is well known they tend to the origin in definite directions (unstable node) when $\lambda \geq 2$, or else spirally (unstable focus) when $\lambda < 2$.

The paths in the phase plane, outside the limit-cycle are indicated in Figure 18.

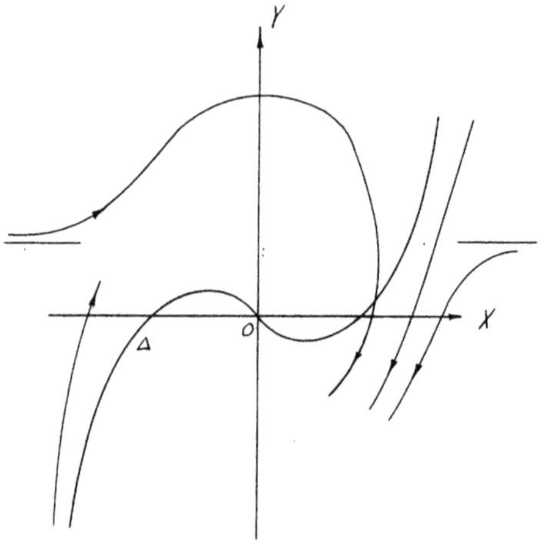

Figure 18

The main result of our discussion may be summarized as follows:

(10.2) THEOREM. Every path in the plane (with the origin omitted) tends toward the limit-cycle.

It may be observed that on the sphere the critical point B behaves like an ordinary saddle point such that the four paths tending to the saddle point are tangent to one another. On the hemisphere of Figure 16 two of the "hyperbolic" angles are shown. The two others are on the other hemisphere. Similarly of course for the point B'.

BIBLIOGRAPHY

[1] POINCARÉ, H., "Sur les courbes définies par les équations différentielles," Oeuvres, Vol. 1, 1-158, 167-222.

[2] BENDIXSON, I., "Sure les courbes définies par les équations différentielles," Acta, 24, 1-88 (1901).

[3] NIEMITZKI, V., and STEPANOV, V., "Qualitative theory of differential equations," GITL, Moscow, 1st Ed. 1947, 2nd ed. 1949.

[4] LEFSCHETZ, S., "Lectures on differential equations," Princeton University Press, 1946.

[5] PICARD, E., "Traité d'Analyse, II , 3rd edition.

[6] LEVINSON, N., "On The Existence of Periodic Solutions for Second Order Differential Equations with a Forcing Term," Journal of Mathematics and Physics, 22, pp. 41-48, 1943.

NOTE: This paper was prepared under contract with the Office of Naval Research, and was equally sponsored by the Office of Air Research.

A METHOD FOR THE CALCULATION OF LIMIT CYCLES BY SUCCESSIVE APPROXIMATION

By John McCarthy

Consider a system of differential equations of the form

$$\frac{dx_i}{dt} = f_i(x_1, \ldots, x_n) \qquad i = 1, \ldots, n$$

which we write in the vector form*

1) $$\frac{dx}{dt} = f(x) .$$

Let $\varphi(t)$ be a periodic solution of 1). Then in the neighborhood of $\varphi(t)$ there exist coordinates $(y_1, \ldots, y_{n-1}, s)$ such that in these coordinates t can be eliminated from 1) which then takes the form

2) $$\frac{dy}{ds} = g(y, s)$$

where $g(y, s)$ has period ω in s. The solution $\varphi(t)$ of 1) goes into a solution $\Psi(s)$ of 1) of period ω.

$\varphi(t)$ is called a stable limit cycle of 1) if the linear differential system with periodic coefficients

$$\frac{du_i}{ds} = \sum_{j=1}^{n-1} \frac{\partial}{\partial y_j} g_i(\Psi_1(s), \ldots, \Psi_{n-1}(s), s)\, u_j$$

or in vector notation

3) $$\frac{du}{ds} = \nabla_y g(\Psi(s), s)\, u = G(s)\, u$$

has all its characteristic exponents with negative real parts. This implies that all solutions passing sufficiently close to $\Psi(s)$ tend toward this solution.

Let $\gamma_n(s)$, where $\gamma_n(s + \omega) = \gamma_n(s)$, be an approximation to $\Psi(s)$. We propose to find a better approximation $\gamma_{n+1}(s)$. Form the linear differential system

4) $$\frac{dy}{ds} = g_n(s) + G_n(s)\, [y - \gamma_n(s)]$$

where $g_n(s) = g(\gamma_n(s), s)$ and $G_n(s) = \nabla_y g(\gamma_n(s), s)$, and let $\gamma_{n+1}(s)$ be the solution of 4) of period ω. We then have

THEOREM 1. Given the differential system 1) and the

---

* In what follows all functions are assumed to be sufficiently differentiable. It will be clear from the proofs although it will not be explicitly stated that any new functions will be sufficiently differentiable. Vector notation is as in [1].

stable limit cycle $\varphi(t)$ there exist constants $_{o}$, $k > 0$ with $\frac{1}{k} \leq \varepsilon_o$ such that if $||\gamma_n(s) - \Psi(s)|| < \varepsilon < \varepsilon_o$ then $\gamma_{n+1}(s)$ is uniquely defined as the periodic solution of 4) and

5) $$||\gamma_{n+1}(s) - \Psi(s)|| < k\,\varepsilon^2.$$

Hence if $||\gamma_o(s) - \Psi(s)|| < \varepsilon_o$, then $\gamma_n(s) \to \Psi(s)$ uniformly as $n \to \infty$.

Before proving the theorem we first prove

LEMMA 1. If the differential system

6) $$\frac{dx}{ds} = A(s)\, x$$

where $A(s + \omega) = A(s)$ is such that the characteristic roots of the matrix $Y(\omega)$, where $Y(s)$ is the solution of the matrix equation $\frac{dy}{ds} = AY$ with $Y(o) = I$, are all different from 1, then there exists a constant $M(A)$ depending continuously on A such that the equation

7) $$\frac{dx}{ds} = A(s)x + b(s)$$

where $b(s + \omega) = b(s)$ has a unique periodic solution $x_o(s)$ of period ω with

8) $$||x_o(s)|| \leq M(A)\; ||b(s)||$$

PROOF. Let $Y(s)$ be the matrix solution of

$$\frac{dY}{ds} = A(s)Y$$

with $Y(0) = I$. Then any solution of 7) may be expressed in the form

$$x_o(s) = Y(t)\left[x(0) + \int_0^s Y^{-1}(s_1)\, b(s_1)\, ds_1\right].$$

$x_o(s)$ is distinguished by the periodicity condition

$$x_o(0) = x_o(\omega) = Y(\omega)\left[x_o(0) + \int_0^\omega Y^{-1}(s_1) b(s_1) ds_1\right]$$

or

$$x_o(0) = \left[I - Y(\omega)\right]^{-1} Y(\omega) \int_0^\omega Y^{-1}(s_1)\, b(s_1)\, ds_1$$

giving

$$||x_o(s)|| \leq ||Y(s)||\; \omega\; ||Y^{-1}(s)||\;\; ||Y(\omega)||\; ||[I - Y(\omega)]^{-1}|| + 1\;\; ||b(s)|| = M(A)\; ||b(s)||.$$

Here $[I - Y(\omega)]^{-1}$ exists because none of the characteristic roots of $Y(\omega)$ are 1. The bound $M(A)$ depends continuously on A because of the theorem of continuous dependence of a solution with fixed initial point on the right hand side of the differential equation.

PROOF OF THEOREM 1. Consider the linear differential system

9) $$\frac{du}{ds} = G_n(s)\, u + \left[g_n(s) + G_n(s)\,\{\Psi(s) - \gamma_n(s)\} - g(s)\right]$$

where $g(s) = g(\Psi(s), s)$. By Taylor's theorem there exist constants a_1, $k_1 > 0$ such that if $\|\gamma_n - \Psi\| < \varepsilon < a_1$ then

10) $$\|g_n + G_n(\Psi - \gamma_n) - g\| < k_1\, \varepsilon^2 .$$

Choose a_2 such that for $\|\gamma_n - \Psi\| < a_2$, $M(G_n) < 2M(G)$, where $M(G)$ is determined from Lemma 1, let $k = 2k_1$, and let $\varepsilon_0 = \min\,(a_1, a_2, \frac{1}{k})$. Then, if $\|\gamma_n - \Psi\| < \varepsilon < \varepsilon_0$, 6) has a unique periodic solution $u_0(s)$ such that $\|u_0(s)\| < k\,\varepsilon^2 < \varepsilon_0$. However, $y = u + \Psi$ satisfies 4) if and only if u satisfies 6). Hence, under the hypothesis of the theorem, 4) has a unique periodic solution $\gamma_{n+1}(s)$ satisfying 5). This proves the theorem.

REMARK. If the successive steps are carried out approximately with an error $0 < \eta < \frac{1}{2}\,\varepsilon_0$ starting within $\frac{1}{2}\,\varepsilon_0$ of $\Psi(s)$, then $\Psi(s)$ will eventually be determined to within an error 2η. That is, the process is stable.

This process involves finding the periodic solution of the linear system with periodic coefficients 4) at each step, which is not a simple matter. However, it can be facilitated by determining at each step the general solution $Y_{n+1}(s)$ of

11) $$\frac{dY_{n+1}}{ds} = G_n\, Y_{n+1}$$

where $Y_{n+1}(s)$ is a matrix solution such that $Y_{n+1}(0) = I$.

$Y_{n+1}(s)$ is to be determined by the Picard iteration scheme

12) $$Y_{n+1,\,k+1}(s) = I + \int_0^s G_n(s_1)\, Y_{n+1,\,k}(s_1)\, ds_1 \quad .$$

$\gamma_{n+1}(s)$ is then determined by the formula

13) $$\gamma_{n+1}(s) = Y_{n+1}(s)\left[\gamma_{n+1}(0) + \int_0^s Y^{-1}_{n+1}(s_1)\,\{g_n(s_1) - G_n(s_1)\gamma_n(s_1)\}\, ds_1\right]$$

where $\gamma_{n+1}(0)$ is determined from the periodicity condition to be

14) $$\gamma_{n+1}(0) = \left[I - Y_{n+1}(\omega)\right]^{-1} Y_{n+1}(\omega) \int_0^\omega Y^{-1}_{n+1}(s)\,\{g_n(s) - G_n(s)\,\gamma_n(s)\}\, ds$$

The inversion of the matrices $Y_{n+1}(s)$ and $I - Y_{n+1}(\omega)$ can be conveniently accomplished by the iterative scheme

15) $$(A^{-1})_{k+1} = 2(A^{-1})_k - (A^{-1})_k\, A(A^{-1})_k$$

where we start with $Y_n^{-1}(s)$ and $\left[I - Y_n(\omega)\right]^{-1}$.

Suppose that a differential system of the form 2), a limit cycle of which we wish to determine, can be imbedded in a 1 parameter family of differential systems

16) $$\frac{dy}{ds} = g(y, s, \lambda) \qquad 0 \leq \lambda \leq 1$$

satisfying the condition

1) $g(y, s + \omega, \lambda) = g(y, s, \lambda)$
2) For each value of λ, $g(y, s, \lambda)$ has a limit cycle $\Psi(s, \lambda)$ whose characteristic exponents have negative real parts, $\Psi(s, \lambda)$ depending continuously on λ.
3) $g(y, s, 1) = g(y, s)$ and $\Psi(s, 1) = \Psi(s)$ the limit cycle we wish to calculate.
4) We know $\Psi(s, 0)$ and also know the general solution of the variational equations of $\Psi(s, 0)$.

It is then clear that we can determine $\Psi(s)$ by dividing the interval $0 \leq \lambda \leq 1$ into segments $0 < \lambda_1 < \lambda_2 < \cdots < \lambda_k < 1$ and using a good approximation to $\Psi(s, \lambda_k)$ as first approximation to $\Psi(s, \lambda_{k+1})$.

REMARKS.

1. In carrying out the procedure one would wish to divide the interval $0 \leq \lambda \leq 1$ into as few parts as possible and use as few iterations as possible for each k until one got to iterating for $\Psi(s, 1)$. However, if one tried to make too big a jump one would get lost.

2. Probably the easiest way to carry out the computations in an actual case is to use trigonometric polynomials of some fixed degree as approximations to the periodic function involved.

3. In carrying out the "homotopy" it might be necessary to change coordinate systems during the calculation in order to keep the functions single valued. See Figure 1.

4. This method can be interpreted in terms of L. V. Kantorovič's extension of Newton's method to function spaces [2].

EXAMPLE: Consider the equation in polar coordinates

18) $$\frac{dr}{d\theta} = r(1 - r) .$$

Here $g(r, \theta) = r(1 - r)$ and $G(r, \theta) = 1 - 2r$.

If we take $\gamma_0(\theta) = r_0$, then $\gamma_1(\theta)$ satisfies

$$\frac{dr}{d\theta} = r_0(1 - r_0) + (1 - 2r_0)(r - r_0)$$

the periodic solution of which is

$$\gamma_1(\theta) = r_1 = r_o + \frac{r_o(1 - r_o)}{2r_o - 1}$$

This process converges to $r = 1$ if $r_o > \frac{1}{2}$.

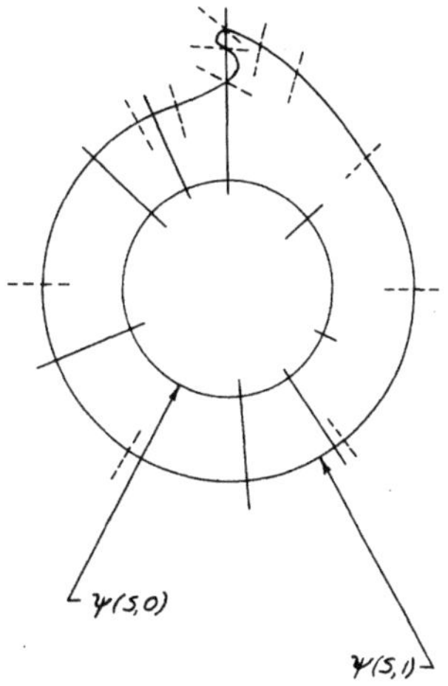

Figure 1

—— s = const. in original system

---- s = const. in second system

BIBLIOGRAPHY

[1] LEFSCHETZ, S., "Lectures on Differential Equations," Annals of Mathematics Studies, Princeton University Press, 1946.

[2] KANTOROVIC, L. V., "On Newton's Method for Functional Equations," Doklady Akad Nauk SSSR (N.S), 59, 1237-1240, 1948. (Russian).

ASYMPTOTIC EXPANSIONS OF SOLUTIONS OF SYSTEMS OF ORDINARY LINEAR DIFFERENTIAL EQUATIONS CONTAINING A PARAMETER

By H. L. Turrittin

§1. INTRODUCTION*

Consider a system of differential equations

$$\varepsilon^h \dot{X} = A(t, \varepsilon) X \tag{1}$$

where h is a non-negative integer; ε is a small parameter; X is a vector

$$X = \begin{Vmatrix} x_1(t, \varepsilon) \\ x_2(t, \varepsilon) \\ \vdots \\ x_N(t, \varepsilon) \end{Vmatrix} \quad ;$$

$\dot{X}$ is the derivative $\frac{dX}{dt}$; and the square matrix $A(t,\varepsilon)$ has an asymptotic expansion

$$A = A(t,\varepsilon) = \sum_{k=0}^{\infty} \varepsilon^k A_k(t)$$

in the domain D_1 where $a \leq t \leq b$ and ε is in a regular region R of the complex ε-plane extending from the circle $|\varepsilon| = \varepsilon_1 > 0$ into the origin $\varepsilon = 0$. All boundary points of R, save the origin $\varepsilon = 0$, belong to R. Trjitzinsky(11) has indicated that an equation of type (1) will possess N independent vector solutions of the form

$$X_i = \left[\sum_{k=0}^{\infty} \sigma^k X_{ik}(t)\right] \exp \{ q_i(t,\varepsilon) \} \qquad (i = 1,\ldots,N) \tag{2}$$

where $\sigma = \varepsilon^{1/r}$; r is a suitable positive integer; the X_{ik}'s are vectors;

$$X_{i0}(t) \not\equiv 0 \ ; \ q_i(t, \ ) = \int_a^t \varepsilon^{-h} \rho_i(t, \varepsilon)\, dt$$

and

$$\rho_i(t,\varepsilon) = \sum_{k=0}^{hr-1} \sigma^k \rho_{ik}(t). \tag{3}$$

---

* This paper was written at Princeton University on sabbatical leave from the University of Minnesota while working under the direction of Prof. S. Lefschetz on Navy Project N6 ori-105, task order V for the Office of Naval Research. The author wishes particularly to thank Prof. Lefschetz for his suggestions and inspiration.

(11) References are listed at the end of this paper.

The $\rho_{ik}(t)$ are roots of certain characteristic equations presently to be introduced. The first objective of this paper is to give a definite procedure for computing these formal solutions. The second objective is to show that these formal solutions are asymptotic solutions in the sense of Poincaré[1]. The asymptotic solutions are then used to solve the non-homogeneous equation corresponding to (1); see §13.

Formal and asymptotic solutions for a single linear differential equation of the Nth order

$$\sum_{i=0}^{N} \varepsilon^{ih} p_i(t,\varepsilon) \frac{d^i x}{dt^i} = p_{-1}(t,\varepsilon) \tag{4}$$

where x is a scalar;

$$p_i(t,\varepsilon) = \sum_{j=0}^{\infty} \varepsilon^j p_{ij}(t), \qquad i = -1, 0, 1, \ldots, N-1;$$

and $p_N(t,\varepsilon) \equiv 1$, as well as for systems of type (1), have been studied by a number of mathematicians, see references (2) - (12). Asymptotic solutions have been extensively used in boundary value problems and more recently in problems relating to degeneracy, see for example Gradstein's[13] work. References for various applications in physics may be found in Jeffreys'[14] book. The type of analysis presented here is also part of the WKB-method used in quantum mechanics. Moreover in the study on nonlinear systems by means of the variational equations one may be led to consider equations of type (1), see for instance the work of Wasow[22] on singular perturbation problems.

To proceed to the detailed analysis, we make the following hypotheses:

(i) All the elements $a_{ij}(t,\varepsilon)$ of the matrix $A(t,\varepsilon)$ in (1) possess derivatives of all orders with respect to t and are analytic in ε in the domain D_1.

(ii) If $a_{ijk}(t)$ is the element in the ith row and j^{th} column of $A_k(t)$, $(i, j = 1, \ldots, N;\ k = 0, 1, 2, \ldots)$, then each $a_{ijk}(t)$ possesses derivatives of all orders in $[a, b]$.

(iii) If $\lambda_i(t)$ and $\lambda_j(t)$ are any two roots of the characteristic equation $|A_0(t) - \lambda I| = 0$; then on $[a, b]$ either $\lambda_i(t) \equiv \lambda_j(t)$ or $\lambda_i(t) \neq \lambda_j(t)$. (The characteristic roots $\lambda_i(t)$ will serve as the $\rho_0(t)$ in (3); $i = 1, \ldots, N$).

(iv) Likewise any two roots of each subsequent auxiliary characteristic equation utilized in our reductions are either identically equal or distinct on $[a, b]$; just as in (iii).

(v) The structure of certain canonical matrices do not change form in the interval $[a, b]$, see §4.

(vi) Certain elements which appear in the matrices derived from the $A_k(t)$ in subsequent reductions are either non-vanishing or identically zero on $[a, b]$, see §8.

In any given problem it usually turns out that in order to satisfy these hypotheses the closed interval $[a, b]$ must be chosen sufficiently short so as to avoid certain special points on the t-axis, the so-called "turning points". The present analysis does not provide information relating to the behavior of solutions when $|\varepsilon|$ is small and t is in the neighborhood of a turning point. Some headway has been made in treating this difficult problem, see for example the recent work of Langer(15,16), Cherry(17), and Evans(18).

A method for computing the formal solutions will now be presented. The analysis is broken down in a sequence of special cases of successively increasing complexity.

§2. CASE I: $A_0 + \varepsilon A_1 + \dots + \varepsilon^{h-1} A_{h-1} \equiv 0$.

If on examination of $\dot{A}$ it is found that A_0 is identically zero, cross out A_0, divide by ε and (1) take the form

$$\varepsilon^{h-1} \dot{X} = (A_1 + \varepsilon A_2 + \dots)X. \tag{5}$$

If $A_1 \equiv 0$, likewise reduce (1) to

$$\varepsilon^{h-2} \dot{X} = (A_2 + \varepsilon A_3 + \dots)X.$$

Continue reducing the exponential on ε until an $\dot{A}_i$ is reached which does not vanish identically. If fortunately we can proceed thus and diminish h to zero, an equation of the form

$$\dot{X} = \sum_{i=0}^{\infty} \varepsilon^i A_i'(t)X \tag{6}$$

is reached which possess N independent vector solutions of the form

$$X = \sum_{j=0}^{\infty} \varepsilon^j X_j(t). \tag{7}$$

To find the $X_j(t)$, $(j = 0, 1, \dots)$, substitute series (7) into (6) and equate successive powers of ε to zero. Thus, again dropping termporarily the t,

$$\begin{aligned} \dot{X}_0 &= A_0'X_0 \\ \dot{X}_1 &= A_0'X_1 + A_1'X_0 \\ &\dots \\ \dot{X}_n &= A_0'X_n + A_1'X_{n-1} + \dots + A_n'X_0 \\ &\dots \end{aligned} \tag{8}$$

The $X_0(t)$, $X_1(t)$, $X_2(t)$, ... are computed in succession by solving these differential equations.

The first equation of (8) has, as is well known, N independent vector solutions which can and will be assembled into a single square matrix solution $\tilde{X}_0(t)$, where each column is one of the independent vector solutions. To make the choice of $\tilde{X}_0(t)$ specific take for $\tilde{X}_0(t)$ the particular solution which satisfies the initial conditions $\tilde{X}_0(a) = I$. Likewise square matrix solutions $\tilde{X}_i(t)$ can be found for the successive equations in (8). In particular take

$$\tilde{X}_n(t) = \tilde{X}_0(t) \int_a^t \tilde{X}_0^{-1}(\tau) \left[A_1'\tilde{X}_{n-1}(\tau) + \dots + A_n'\tilde{X}_0(\tau)\right] d\tau .$$

Then if

$$A_0 + \varepsilon A_1 + \dots + \varepsilon^{h-1} A_{h-1} \equiv 0 , \tag{9}$$

the square matrix $\tilde{X} = \sum_{i=0}^{\infty} \varepsilon^i \tilde{X}_i(t)$ exhibits N independent formal solutions of (1), for the lead matrix $\tilde{X}_0$ is non-singular.

This procedure will be blocked at a certain stage if one of the A's in (9) is not identically zero. We are forced therefore to examine the situation when at least some one element of A_0 is not identically zero and begin by considering a system of the first order.

§3. CASE II: THE FIRST ORDER EQUATION

If $N = 1$, equation (1) degenerates into a single equation

$$\varepsilon^h \dot{x} = \sum_{i=0}^{\infty} \varepsilon^i A_i(t)x \tag{10}$$

in one unknown $x(t, \varepsilon)$ where the A_i's are scalars.

Set $x = y \exp \left\{ \int_a^t \varepsilon^{-h} \rho(\tau, \varepsilon)d\tau \right\}$ in (10) with

$$\rho(t, \varepsilon) = \sum_{i=0}^{h-1} \varepsilon^i A_i(t).$$

Then

$$\varepsilon^h \left[\dot{y} + \varepsilon^{-h} \rho(t, \varepsilon)y\right] = Ay$$

or

$$\varepsilon^h \dot{y} = \left[\varepsilon^h A_h + \varepsilon^{h+1} A_{h+1} + \dots\right] y.$$

On dividing out ε^h, (10) is reduced to an equation of type (6). In this way the desired single independent (i.e., non-trivial) solution is procured by the procedure in Case I regardless of how large h may be.

§4. CASE III: ALL ROOTS EQUAL AND NO 1'S IN CANONICAL FORM

If an element of A_0 is not zero, $N > 1$, and $h \geq 1$, normalize A_0; i.e., make a non-singular linear normalizing transformation

$$X = P(t)Y \tag{11}$$

so that (1) becomes

$$\varepsilon^h(\dot{P}Y + P\dot{Y}) = APY$$

or

$$\varepsilon^h \dot{Y} = (P^{-1}AP - \varepsilon^h P^{-1}\dot{P})Y . \tag{12}$$

The non-singular matrix $P(t)$ can and will be chosen so that $P^{-1}A_0P$ is in canonical form; i.e., so that

$$P^{-1}A_0P = \begin{Vmatrix} M_1 & 0 & \cdots & 0 \\ 0 & M_2 & \ddots & \vdots \\ \vdots & & \ddots & 0 \\ 0 & \cdots & 0 & M_n \end{Vmatrix} = B_0 \tag{13}$$

where all elements are zero save those in the diagonal matrices M_i. Each M_i has the structure

$$M_i = \begin{Vmatrix} \rho_i(t) & 0 & \cdots & \\ \beta_i & \rho_i(t) & & \\ 0 & \beta_i & \ddots & \\ \vdots & & \ddots & \\ & & \beta_i & \rho_i(t) \end{Vmatrix} \tag{14}$$

where $\rho_i(t)$ is a root of the characteristic equation $|A_0 - \lambda I| = 0$ and β_i is either zero or one; all other elements in M_i are zero.

Strictly such a reduction to canonical form is possible for each particular value of t; but for certain special values of t the canonical forms may suddenly change; for example the number and location of the 1's in the M_i matrices may change. It is presumed that at the outset the interval $[a, b]$ has been chosen short enough so as to avoid such changes. This is hypothesis (v), see §1.

At this point we are dealing with an equation (12) of the form

$$\varepsilon^h\dot{Y} = (B_0 + \varepsilon B_1 + \ldots)Y \quad , \quad h \geq 1, \tag{15}$$

where B_0 has the canonical form (13) and at least some element in B_0 is not identically zero. All the elements $b_{ijk}(t)$ of each matrix B_k possess derivatives of all orders on $[a, b]$, just as was the case for the $a_{ijk}(t)$. As special case III assume that $n = 1$, (i.e. there is only one characteristic root $\rho_1(t)$ of multiplicity $N > 1$), and that $\beta_1 = 0$. In this event use on (15) an exponential transformation

(16) $$Y = Z\exp \int_a^t \varepsilon^{-h} \rho_1(t)dt$$

and then the differential system becomes on dividing out an ε

$$\varepsilon^{h-1}\dot{Z} = (B_1 + \varepsilon B_2 + \dots)Z .$$

The process as outlined so far in case I and III is reapplied to this equation. Eventually in a finite number of steps the desired formal solutions are procured provided that at no stage is an equation of type (15) encountered where two or more distinct roots ρ_i appear or where just one root ρ appears with at least one of the β's equal to 1. Granting that this procedure can be carried through, at the last stage when $h = 0$, N formal independent solutions of an equation

$$\dot{W} = (C_0 + \varepsilon C_1 + \dots)W$$

are found as in case I and exhibited in a square matrix $\tilde{W} = \tilde{W}_0 + \varepsilon\tilde{W}_1 + \dots$ with the determinant $|\tilde{W}_0| \neq 0$. In working back from $\tilde{W}$ through a sequence of transformations of types (11) and (16) to a square matrix solution $\tilde{X}$ of the original equation (1), it is clear that $\tilde{X}$ will have the prescribed form (2), and that the columns of $\tilde{X}$ represent independent formal vector solutions, for the transformations of types (11) and (16) are all non-singular.

If this procedure can not be carried through, proceed to the next case.

§5. CASE IV: ALL ROOTS DISTINCT, $h \geq 1$, $N > 1$, $n > 1$

Reconsider equation (15) and assume that the distinct characteristic roots of $|A_0 - \lambda Y| = 0$ are respectively $\rho_1(t), \dots, \rho_K(t)$; $E > 1$; and that the respective multiplicities of these roots are $m_1, \dots, m_K$ with $m_1 + m_2 + \dots + m_K = N$. It will now be shown by utilizing transformations of the form

$$Y = (I + \varepsilon^k Q_k)Z, \quad k = 1, 2, \dots,$$

that system (15) of order N can be decomposed into K separate systems with respective orders $m_1, \dots, m_K$, each lower than N.

Utilizing (17), (15) becomes

$$\varepsilon^h\left[\varepsilon^k\dot{Q}_k Z + (I + \varepsilon^k Q_k)\dot{Z}\right] = (B_0 + \varepsilon B_1 + \dots)(I + \varepsilon^k Q_k)Z$$

or

$$\varepsilon^h\dot{Z} = (I + \varepsilon^k Q_k)^{-1} \left[(B_0 + \varepsilon B_1 + \dots)(I + \varepsilon^k Q_k) - \varepsilon^{h+k} \dot{Q}_k\right] Z.$$

Noting that

$$\left[I + \varepsilon^k Q_k\right]^{-1} = I - \varepsilon^k Q_k + \varepsilon^{2k} Q_k^2 \dots$$

and letting the three dots take care of all terms of greater order than ε^k in ε,

$$\varepsilon^h\dot{Z} = \{B_o + \varepsilon B_1 + \ldots + \varepsilon^{k-1} B_{k-1} + \varepsilon^k C_k + \ldots\}Z$$

where

(18) $$C_k = B_k + B_o Q_k - Q_k B_o.$$

The matrices C_k, B_k, and Q_k can be subdivided into smaller blocks in just the same way as the normalized matrix B_o, see (13). After this subdivision is made, denote the block in the rth row of blocks and s^{th} column respectively by C_{rs}, B_{rs}, Q_{rs} and the elements in the ith row and j^{th} column of these rectangular matrices respectively by c_{ij}, b_{ij}, q_{ij}, with $i = 1, \ldots, R$ and $j = 1, \ldots, S$. <u>A judicious choice of q_{ij} can and will be made so that all the c_{ij} in C_{rs} are identically zero,</u> $i = 1, \ldots, R$; $j = 1, \ldots, S$, provided $\rho_r(t) \neq \rho_s(t)$.

To see this first note that the diagonal matrix M_r in (14) can be written in the form

$$M_r = \rho_r I + E_r$$

where I is the unit matrix and E_r is a square matrix made up of elements β_r running down the secondary diagonal, just below the main diagonal, with zeros elsewhere. Likewise

(19) $$M_s = \rho_s I + E_s$$

From (18)

$$C_{rs} = B_{rs} + M_r Q_{rs} - Q_{rs} M_s = B_{rs} + (\rho_r I + E_r)Q_{rs} - Q_{rs}(\rho_s I + E_s)$$

or

$$c_{ij} = b_{ij} + (\rho_r - \rho_s)q_{ij} + \beta_r q_{i-1,j} - \beta_s q_{i,j+1}$$

where

$$q_{o,j} = 0; \quad j = 1, \ldots, S; \quad \text{and} \quad q_{i,S+1} = 0; \quad i = 1, \ldots, R.$$

The q_{ij} which make the c_{ij} zero are therefore determined by the system

$$(\rho_s - \rho_r)q_{ij} + \beta_s q_{i,j+1} - \beta_r q_{i-1,j} = b_{ij}; \quad i = 1, \ldots, R; \quad j = 1, \ldots, S.$$

Solving

$$q_{1S} = b_{1S} \,/\, (\rho_s - \rho_r)$$

and

$$q_{i+1,S} = (b_{i+1,S} + \beta_r q_{i,S}) \,/\, (\rho_s - \rho_r), \quad i = 1, \ldots, R-1.$$

Thus the elements q_{ij} in the last column of Q_{rs} can be calculated by working from the top down. Next working from right to left the

q_{ij}'s in successive columns can be evaluated. In particular if the q_{ij}'s in the $(j+1)^{st}$ column are known, $j = 1, \ldots, S-1$, then the q's in the jth column, working from the top down, are given in succession by

$$q_{1j} = (b_{1j} - \rho_s q_{1,j+1}) \,/\, (\rho_s - \rho_r)$$

and

$$q_{i+1,j} = (b_{i+1,j} + \rho_r q_{i,j} - \rho_s q_{i+1,j+1}) \,/\, (\rho_s - \rho_r); \quad i = 1, \ldots, R-1.$$

By choosing the q_{ij} in this way all C_{rs} blocks which correspond to distinct roots $\rho_r(t)$ and $\rho_s(t)$ are filled with zeros. This is done first using a transformation (17) with $k = 1$, then $k = 2, 3, \ldots$ and so on to infinity. Set all q_{ij}'s in blocks where $\rho_r(t) \equiv \rho_s(t)$ equal to zero. Thus a non-singular formal zero-inducing transformation

$$Y = \left[(I + \varepsilon Q_1)(I + \varepsilon^2 Q_2)(I + \varepsilon^3 Q_3)\cdots\right] Z \tag{21}$$

has been found which reduces (15) to

$$\varepsilon^h \dot{Z} = CZ \tag{22}$$

where

$$C = \left\| \begin{matrix} G_1 & 0 & \cdots & 0 \\ 0 & G_2 & & \vdots \\ \vdots & & \ddots & \vdots \\ 0 & \cdots & \cdots & G_K \end{matrix} \right\| . \tag{23}$$

In (23) the diagonal matrix G_i is of order m_i, $(i = 1, \ldots, K)$. Each lead matrix G_{io} in the expansion $G_i = G_{io} + \varepsilon G_{i1} + \varepsilon^2 G_{i2} + \cdots$ is itself in a canonical form of type (13) with the same root ρ_i appearing in each diagonal block of G_{io}.

Substitution (21) is merely a formal transformation because in general the infinite product in (21) may be divergent. It is at this stage that the need for derivatives of all orders on the elements $b_{ijk}(t)$ (and therefore on the $a_{ijk}(t)$'s) comes into play. The existence of each successive Q_k requires the existence of successively higher ordered derivatives on the $b_{ijk}(t)$'s and on the $a_{ijk}(t)$'s. Note also that derivatives of all orders exist for the elements in each of the G_{ij} matrices.

System (22) can therefore be split into K separate systems

$$\varepsilon^h \dot{Z}_i = G_i Z_i \,, \qquad i = 1, \ldots, K, \tag{24}$$

each of lower order than N. Once m_i independent vector solutions of the prescribed form (2) have been found for each system (24), the solutions for each system are assembled into a square matrix $\tilde{Z}_i$. Then

$$\tilde{Z} = \left\| \begin{matrix} \tilde{Z}_1 & 0 & \cdots & 0 \\ 0 & \tilde{Z}_2 & & \\ \vdots & & \ddots & \\ 0 & & & \tilde{Z}_K \end{matrix} \right\|$$

exhibits N independent solutions of (22) and via non-singular transformations of type (21), (11), and (16) we get back to the desired solution $\tilde{X}$ exhibiting N independent formal solutions of (1).

If all the characteristic roots are distinct, as will be assumed in case IV, system (1) at this stage splits into N separate equations all of order one and each of the type considered in case II. Therefore, if all the characteristic roots are distinct N formal independent solutions (2) of (1) can be found.

§6. CASE V: A SINGLE ROOT OF MULTIPLICITY m, $\mu_0 \geq 1$

In the event that all the characteristic roots are not distinct, proceed as in case IV; decompose the system into several separate systems (24) of lower order; and focus attention on any particular one of these systems, say on one of the form

$$\varepsilon^h \dot{Y} = (B_0 + \varepsilon B_1 + \ldots)Y$$

where B_0 has the canonical structure of G_{10}, i.e., with a single root ρ running down the entire main diagonal. Assume also that at least one 1 appears on the secondary diagonal.

Furthermore, when the normalization (11) is made P can be chosen so that in B_0 either the β_i are all 1's, (see (14)); or β_1 is zero and the remaining β_i are 1's. Also in each normalization P can and will be so chosen that the diagonal blocks M_1, M_2, ..., which contain β_i's equal to 1, are arranged in order of size as they run down the diagonal, the largest (if there is one) at the bottom.

Again transformations of type (17) are utilized to annul certain elements in the matrix coefficients of the powers of ε. In particular the equations which control the choice of the q_{ij} become

$$(25) \qquad \beta_s q_{i,j+1} - \beta_r q_{i-1,j} = b_{ij};\ i = 1, \ldots, R;\ j = 1, \ldots, S;$$

for in this case $\rho_r = \rho_s$, see (20). It is not necessary for our purpose, nor is it possible, to annul all the elements in C_{rs}, (see §5). Certain of the q_{ij}'s in Q_k will therefore remain uncalculated; set all such q_{ij}'s equal to zero.

Consider a block C_{1s}, $s > 1$, then if $\beta_1 = 0$, (25) becomes $q_{i,j+1} = b_{ij}$; $i = 1, \ldots, R$; $j = 1, \ldots, S - 1$; and hence with this choice of the q's every element of C_{1s} is made zero except those in the last column.

Likewise, for a block C_{rs}, $r > 1$, $s > 1$, and $s \geq r$, (25) becomes $q_{1,j+1} = b_{1j}$; $j = 1, \ldots, S-1$; and

$$q_{i,j+1} = b_{ij} + q_{i-1,j}; \quad i = 2, \ldots, R; \quad j = 1, \ldots, S-1$$

where $q_{i1} = 0$ for $i = 1, \ldots, R$. This choice of q's again annuls every element in C_{rs} save those in the last column.

In this fashion zeros are induced into the indicated C_{rs} by transformations of type (17) first with $k = 1$, then with $k = 2, 3, \ldots$ to infinity. For example, if we start with an equation (15) in which

$$(26) \qquad B_0 = \left\| \begin{array}{cc|cc|ccc} \rho & & & & & & \\ & \rho & & & & & \\ \hline & & \rho & & & & \\ & & 1 & \rho & & & \\ \hline & & & & \rho & & \\ & & & & 1 & \rho & \\ & & & & & 1 & \rho \end{array} \right\| = C_0$$

where all missing elements are zero and use transformation (21), (15) is reduced to

$$(27) \qquad \varepsilon^h \dot{Z} = \sum_{i=0}^{\infty} \varepsilon^i C_i Z$$

where $C_0 = B_0$ and

$$(28) \qquad C_i = \left\| \begin{array}{cc|cc|ccc} x & x & 0 & x & 0 & 0 & x \\ x & x & 0 & x & 0 & 0 & x \\ \hline x & x & 0 & x & 0 & 0 & x \\ x & x & 0 & x & 0 & 0 & x \\ \hline x & x & x & x & 0 & 0 & x \\ x & x & x & x & 0 & 0 & x \\ x & x & x & x & 0 & 0 & x \end{array} \right\|, \quad i = 1, 2, \ldots$$

Here x indicates an element which may not be zero. Note also that the elements in every matrix C_i possess derivatives of all orders on $[a, b]$.

Returning to the general problem, when equation (27) is reached set

$$(29) \qquad Z = W \exp \left\{ \int_a^t \varepsilon^{-h} \rho(t) dt \right\}$$

and (27) is transformed into

$$\varepsilon^h (\dot{W} + \varepsilon^{-h} \rho W) = \sum_{i=0}^{\infty} \varepsilon^i C_i W$$

or

$$(30) \qquad \varepsilon^h \dot{W} = [(C_0 - \rho I) + \varepsilon C_1 + \varepsilon^2 C_2 + \ldots] W = DW = \| D_{ij} \| W$$

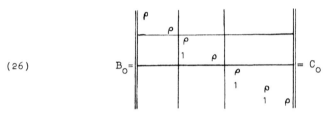

where the D's are introduced as an abbreviation. The effect of transformation (29) has been to erase the ρ running down the main diagonal in C_o without any other change in (27).

This being done and denoting the order of D in (30) by m make the <u>non-singular shearing transformation</u>.

$$(31) \qquad W = \begin{Vmatrix} \varepsilon^{(m-1)\mu} & 0 & . & . & . & 0 \\ 0 & \varepsilon^{(m-2)\mu} & & & & \\ . & & \ddots & & & . \\ . & & & \varepsilon^{2\mu} & & . \\ . & & & & \varepsilon^{\mu} & 0 \\ 0 & . & . & . & 0 & 1 \end{Vmatrix} V$$

and (30) takes the form

$$(32) \qquad \varepsilon^{h}\dot{V} = \| \varepsilon^{(i-j)\mu} D_{ij} \| \, V.$$

Essentially the purpose of transformation (31) is to maintain the general form of equation (30), but to modify the leading coefficient $(C_o - I\rho)$ so that not only are the 1's present on the secondary diagonal in this matrix coefficient, but also other non-zero elements are induced into the lead coefficient <u>above</u> the main diagonal.

To select the proper μ, look at (32) and let μ start at zero and <u>continuously increase</u>. Note first that an element on the main diagonal is unaffected as μ increases. All elements below the main diagonal are multiplied by higher and higher powers of ε as μ goes up; those to the right of the main diagonal are multiplied by lower and lower powers of ε. At the outset each element D_{ij} which is not identically zero can be expanded as a series

$$D_{ij} = \varepsilon^{h_{ij}} \sum_{k=0}^{\infty} \varepsilon^{k} d_{ijk}$$

where $d_{ijo} \neq 0$ and each positive integer h_{ij} is <u>at least one</u>, except for those <u>special</u> elements on the secondary diagonal whose expansion begins with 1. For these special elements the corresponding $h_{ij} = 0$. By hypothesis at least one such special element is present. At the outset, i.e. when $\mu = 0$, the h_{ij} for the special elements are lower in value than those pertaining to other elements which are not identically zero; but after the transformation (31) the respective expansions of the elements begin with the powers $\varepsilon^{(i-j)\mu+h_{ij}}$, excluding from consideration elements which are identically zero. In particular the expansions for the special elements begin with the power ε^{μ}. For sufficiently small μ, this power μ of ε will be less than $(i - j)\mu + h_{ij}$, the power for an ordinary element; but as μ increases a stage will be reached when for the first time $\mu = (i - j)\mu + h_{ij}$ for some i and j, $j \geq i$, for one or more elements if there exists at least one such h_{ij}. Note this critical value μ_o of μ. A special case arises if all elements on and above the main diagonal are identically zero. In this case set $\mu = h$ and (32) is at once reduced to an equation of type (6) with its corresponding formal independent solutions.

If this special case does not occur and μ_0 is 1 or greater, set $\mu = 1$ in (31) and as a consequence in (32) the expansion of each element in the matrix on the right begins with at least the first power of ε, if not a higher power. Therefore, remove a common factor ε from both sides of the equation. Equation (32) then takes the form (5) and we are ready to repeat the entire process as described up to this point. If in fact at each repetition one fails to solve the problem by the methods described in case I - IV and each time reaches a stage as described where $\mu_0 \geq 1$, h is lowered again and again, as just indicated, and in a finite number of stages an equation of type (7) is reached which yields m independent solutions. Again by back tracking through non-singular transformations the desired solution $\tilde{X}$ is reached.

If $\mu_0 < 1$ a more complicated situation arises

§7. CASE VI: $0 < \mu_0 < 1$

If $\mu_0 < 1$, μ_0 is a fraction say $\mu_0 = q/p$ where q and p are positive integers, $q < p$, and q is prime to p. In this case set $\mu = \mu_0$ in (31) and (30) is reduced to (32). Fractional exponents of ε have appeared of necessity for the first time. These fractional exponents are removed at once by introducing a new parameter $\sigma = \varepsilon^{1/p}$ into (32). Then because the expansion of each element in the matrix of (32) begins with at least the qth power of σ, if not a higher power, a common factor of σ^q is divided out of both sides of the equation and (32) takes the form

$$\sigma^h \dot{V} = \sum_{i=0}^{\infty} \sigma^i D_i(t)V$$

where H = ph - q. This equation is precisely of type (1), but this time the exponent h has <u>increased</u> instead of decreased. Despite this the procedure as outlined is reapplied. If on normalizing D_0 in (33), <u>two</u> or <u>more</u> <u>distinct</u> <u>roots</u> are found, the system is split up into two or more distinct systems of type (1) <u>each</u> <u>of</u> <u>lower</u> <u>order</u> <u>than</u> (33). The procedure is again applied to each of these new systems. Either the desired formal solutions are found or new equations of type (33) of still lower order are reached with an even larger H. If the characteristic equations of the new D_0's always yield at least two distinct roots finally we shall reach systems of order 1 with possibly very large H's. These are handled as in case II and the desired independent formal solutions are thus procured.

The only thing which could possibly block this process would be to reach an equation of type (33) where all the characteristic roots of D_0 are alike. This brings us to the last case.

§8. CASE VII: ROOTS OF D_0 ALL ALIKE

In this section it will be shown that if the process outlined in case I - VI is repeatedly carried out it must necessarily terminate after a finite number of stages by yielding the desired formal solutions.

It must be shown first however that, if the roots of D_0 are all alike, this root is identically zero. Observe first that the matrix $(D_0 - \lambda I)$ has a very special structure arising from the fact that (33) was derived from (30) by means of (31). In (30) the $(C_0 - \rho I)$ was a matrix of order m in normal form with a single characteristic root zero of multiplicity m, see (26) with $\rho = 0$, and each C_i contained a number of zeros, see (28). More particularly divide $(C_0 - \rho I)$ into blocks such that

$$C_0 - \rho I = \left\| \begin{array}{c|c|c|c|c} 0 & 0 & \cdots & & 0 \\ \hline 0 & E_2 & & & \\ \hline \vdots & & E_3 & & \vdots \\ \hline & & & \ddots & \\ \hline 0 & \cdots & & & E_L \end{array} \right\|, \quad E_1 = 0,$$

where each E_i, $(i = 2, \ldots, L)$, is a matrix of the type used in (19) containing 1's down the secondary diagonal and zeros elsewhere. Also subdivide $(D_0 - \lambda I)$ into corresponding blocks so that

$$D_0 - \lambda I = \left\| \begin{array}{ccc} S_{11} & \cdots & S_{1L} \\ \vdots & & \vdots \\ S_{L1} & \cdots & S_{LL} \end{array} \right\|$$

Then if $\beta_1 = 0$ the effect of transformation (31) is to make

$$(34) \qquad S_{11} = \left\| \begin{array}{cccccc} -\lambda & x & x & \cdots & & x \\ 0 & -\lambda & x & & & \vdots \\ 0 & 0 & -\lambda & & & \vdots \\ \vdots & & & \ddots & & x \\ \vdots & & & & \ddots & \\ 0 & & \cdots & & 0 & -\lambda \end{array} \right\|$$

where all elements below the diagonal are zero and where those above the diagonal may not be identially zero. Here and in subsequent matrices an x indicates an element which may not be identially zero. If $\beta_1 = 1$, S_{11} will have the structure indicated in (35).

All the S_{ij} below the main diagonal are identically zero. If the block is above the main diagonal.

$$S_{ij} = \begin{Vmatrix} 0 & 0 & \cdots & 0 & x \\ 0 & & \cdots & 0 & x \\ & & \cdots & & \\ 0 & & \cdots & 0 & x \end{Vmatrix}, \quad j > i.$$

Each diagonal has the structure

$$S_{ii} = \begin{Vmatrix} -\lambda & 0 & \cdots & & x_1 \\ 1 & -\lambda & 0 & & x_2 \\ 0 & 1 & & & \\ & & & -\lambda & x_p \\ 0 & \cdots & 0 & 1 & -\lambda \end{Vmatrix}, \quad i = 2, \ldots, L.$$

It will presently turn out that even $x_1, \ldots, x_p$ are all identically zero.

Because of its peculiar structure the value of the determinant $|D_0 - \lambda I|$ is equal to the product of the determinants of the diagonal blocks S_{ii}. If $\rho_1 = 0$, a matrix S_{11} of type (34) is present and its determinant is equal to some power of λ. Hence $|D_0 - \lambda I| = 0$ has in this case at least one identically zero root. If all roots are equal, the other necessarily must also be identically zero.

If $\rho_1 = 1$, such a block as (34) is missing and all diagonal blocks have the structure (35). By hypothesis at least one such block is present. It can be shown by induction that the determinant

$$(36) \qquad |S_{ii}| = (-1)^{p+1} \left[\lambda^{p+1} - x_p \lambda^{p-1} - x_{p-1} \lambda^{p-2} - \cdots - x_1\right], \quad i > 1.$$

If all the roots of $|D_0 - \lambda I| = 0$ are equal and the root is α, the bracket in (36) would of necessity equal

$$(37) \qquad (\lambda - \alpha)^{p+1} = \lambda^{p+1} - (p + 1) \lambda^p \alpha + \frac{(p + 1)p \lambda^{p-1}\alpha^2}{2} - \cdots .$$

Comparing (36) and (37) it is obvious α must be identically zero. Hence, if all the characteristic roots of $|D_0 - \lambda I| = 0$ are alike, not only are they identically zero as we have just seen, but also every x_j, $(j = 1, \ldots, p)$ in (36) must be identically zero for $i = 2, \ldots, L$.

To see more precisely what transformation (31) accomplishes when the characteristic roots are all zero both before and after (31) is applied, the invariant factors of the two matrices $T_1 = [(C_0 - \rho I) - \lambda I]$ and $T_2 = [D_0 - \lambda I]$ must be compared. Suppose that the diagonal blocks E_1, E_2, $\ldots$, E_L are of order $a_1, a_2, \ldots, a_L$ respectively. Since the order of C_0 is m, $m = a_1 + \cdots + a_L$. The normalization has been so performed that $a_2 \leq a_3 \leq \cdots \leq a_L$. The invariant factors of T_1 are

$$(38) \qquad 1, 1, \ldots, 1, \lambda, \lambda, \ldots, \lambda, \lambda^{a_2}, \lambda^{a_3}, \ldots, \lambda^{a_1}, \ldots, \lambda^{a_j}, \ldots, \lambda^{a_L}$$

where $(m - a_1 - L + 1)$ ones appear on the list and a_1 of the λ's. The greatest common divisor of all 1st order determinants, all 2nd order determinants, etc. formed from T_1 are respectively

$$(39) \qquad 1,\ 1,\ \ldots,\ 1,\ \lambda,\ \lambda^2,\ \ldots,\ \lambda^{a_1},\ \lambda^{a_1+a_2},\ \lambda^{a_1+a_2+a_3},\ \ldots,\ \lambda^m$$

where again $(m - a_1 - L + 1)$ ones appear on the list.

We come now to the final crucial point. Transformation (31) not only converts the characteristic matrix T_1 into T_2, but when the list of greatest common divisors for the determinants of the various orders selected from T_2 is made and this new list is compared with (39) there will be at least as many 1's in it; the new list will terminate as before in λ^m, and <u>the exponents on intermediate λ's will not have increased, and at least one of these exponents will have decreased by at least a unit</u>.

To demonstrate this recall that by hypothesis in T_2 there is at least one element, say x_0, above the main diagonal which is not identically zero. It is desirable to assume at this stage, as well as at each successive shearing, that not only is an $x_0(t)$ not identially zero, <u>but also that</u> $x_0(t) \neq 0$ on $[a, b]$, (hypothesis vi, §1). If this non-zero element is in S_{11} and $\beta_1 = 0$, form a determinant from T_2 by crossing out all rows and columns except those passing through x_0 and the 1's on the secondary diagonal. The value of this determinant is x_0 and the determinant is of order $(m - a_1 - L + 2)$. Since $x_0 \neq 0$ and not divisible by λ, on the new list for T_2 corresponding to (37) the first λ must therefore be replaced by a 1, i.e., the exponent on this λ is decreased from 1 to 0.

If the x_0 had appeared in the right-hand column of a S_{1i}, $i = 2, \ldots, L$, and $\beta_1 = 0$, we would have proceeded in the same way and found the exponent on the first λ would have to be lowered to form a new list of divisors from the old list. Finally let one of the x's in

$$(40) \qquad S_{ij} = \begin{Vmatrix} 0 & 0 & \ldots & 0 & x_1 \\ 0 & 0 & \ldots & 0 & x_2 \\ & & \ldots & & \\ 0 & 0 & \ldots & 0 & x_\alpha \end{Vmatrix},\ a = a_i,\ j > i,\ i > 1.$$

be non-zero, (hypothesis vi).

If λ^i is the greatest common divisor of all determinants formed from T_1 of order η, $(\eta = 1, \ldots, m)$, it is possible to select from T_2 a determinant of order η which will have λ^i as its value. For example, form a determinant from T_2 by crossing out the rows

$$(41) \qquad (a_1 + a_2 + \ldots + a_i + 1)^{th},\ (a_1 + a_2 + \ldots + a_{i+1} + 1)^{th},\ \ldots,\ (a_1 + a_2 + \ldots + a_{L-1} + 1)^{th}$$

and the columns

(42) $(a_1 + a_2 + \dots + a_{i+1})^{th}$, $(a_1 + a_2 + \dots + a_{i+2})^{th}$,

$\dots, (a_1 + a_2 + \dots + a_L)^{th}$.

This determinant is of order $(m - L + i)$ and has the value $\lambda^{a_1+a_2+\dots+a_i}$. This value $\lambda^{a_1+\dots+a_i}$ is the greatest common divisor of all $(m - L + i)^{th}$ ordered determinants of T_1, as may be seen by looking at the list (39).

However, if one of the x-elements in (40) is non-zero, there is also another determinant of order $(m - L + i)$ formed from T_2 not containing $\lambda^{a_1+\dots+a_i}$ as a factor. To construct this determinant cross out the same rows and columns as listed in (41) and (42) except in place of crossing out the column $(a_1 + a_2 + \dots + a_j)^{th}$, retain it, and instead cross out the $(a_1 + a_2 + \dots + a_i)^{th}$ column. As may be verified by induction this determinant has the value

$$(-1)^{a+1} \lambda^{a_1+\dots+a_{i-1}} \left[x_a \lambda^{a-1} + x_{a-1} \lambda^{a-2} + \dots + x_1 \right], \quad a = a_i,$$

where at least one of the x's is not zero. Hence, the greatest common divisor of all $(m - L + i)^{th}$ ordered determinant is some power of λ less than $\lambda^{a_1+\dots+a_i}$.

Hence, if the process of shearing, increasing h, normalizing, and introducing zeros is carried out a finite number of times either the formal solutions will be found by the methods previously described in cases I - VI or the list corresponding to (39) will come to contain all 1's save a final λ^m. This means that at this stage the lead matrix corresponding to $(C_0 - \rho I)$ of (30) would consist of all zeros save for the secondary diagonal which would be filled with 1's. Then when once zeros are induced into the matrices by (21) and the shearing transformation (31) is once again carried out, the lead matrix takes the form (35) and must of necessity have at least two distinct roots if any one of the x's in (35) is not zero with the subsequent decomposition of the problem into two or more systems of lower order; or, if all the x's of (35) are identically zero, the μ in the shearing transformation instead of being set equal to one can be set equal to h (or H) and the system reduced at once to one of type (6) where h (or H) equals zero. Finally, then on repetition a finite number of times either the h (or H) is reduced to zero and the desired independent formal solutions are found or the splitting of the system has occurred often enough to produce first order equations with their accompanying independent solutions.

These facts are summarized in

THEOREM 1. If the hypotheses i-vi in §1 are satisfied, equation (1) has N independent formal vector solutions of the form (2). These formal solutions can be computed by the procedure outlined in cases I-VII and assembled into a matrix of the form

(43) $$\tilde{X}(t, \varepsilon) = \mathcal{X}(t, \varepsilon)E(t, \varepsilon),$$

where

$$\chi(t, \varepsilon) = \sum_{k=0}^{\infty} \sigma^k \tilde{X}_k(t);$$

$\sigma = \varepsilon^{1/r}$ and r is a suitable positive integer;

$$E(t, \varepsilon) = ||\delta_{ij} \exp \{q_j(t, \varepsilon)\}||;$$

δ_{ij} is the Kronecker delta;

(44) $$q_j(t, \varepsilon) = \int_a^t \varepsilon^{-h} \rho_j(\tau, \varepsilon)d\tau \quad ;$$

and

$$\rho_j(t, \varepsilon) = \sum_{k=0}^{hr-1} \sigma^k \rho_{jk}(t) \qquad j = 1, \ldots, N.$$

Each element of every $\tilde{X}_j(t)$ and all $\rho_{jk}(t)$ possess derivatives of all orders on $[a, b]$. In the formal expansion of the determinant

(45) $$|\chi(t, \varepsilon)| = \sigma^p \sum_{k=0}^{\infty} \sigma^k x_k(t)$$

the p is a suitable non-negative integer and $x_0(t) \neq 0$ on $[a, b]$.

§9. ASYMPTOTIC APPROXIMATION TO m TERMS

In this section it will be shown that if the formal solutions are cut short, the truncated series represent asymptotically N independent actual solutions of (1) to a finite number of terms. The analysis will parallel that given by Birkhoff and Langer[10].

At the outset note that on certain curves in the ε-plane the real part

$$R\{\varepsilon^{-h} \rho_i(t, \varepsilon)\} = R\{\varepsilon^{-h} \rho_j(t, \varepsilon)\}.$$

Denote these curves which extend into the origin by B^t_{ij}. If for particular values of i and j the $\rho_i(t, \varepsilon) \equiv \rho_j(t, \varepsilon)$, there are no B^t_{ij} curves for these particular values of i and j. As pointed out by Trjitzinsky[20] the B^t_{ij} curves sufficiently near the origin are simple and at $\varepsilon = 0$ they possess limiting directions. A particular curve will depend on t. As t varies on $[a, b]$ this curve may vary; the angle of the sector within which this variation takes place can be made as small as desired by suitably shortening the t interval.

HYPOTHESIS VII:

When T is in $[a, b]$ there are no B^t_{ij} curves interior to the region R in the ε-plane.

To satisfy this hypothesis it may be necessary to trim down the size of the interval [a, b] and the original region R without cutting R off from the origin. If this trimming is necessary, we presume that it has been done at the outset. Then without loss of generality assume that the subscripts have been so assigned that when t is in [a, b] and ε is in R

$$\rho_{\tau_{j-1}+1}(t, \varepsilon) \equiv \rho_{\tau_{j-1}+2}(t, \varepsilon) \equiv \ldots \equiv \rho_{\tau_j}(t, \varepsilon); \quad j = 1, \ldots, n,$$

where $\tau_0 = 0 < \tau_1 < \tau_2 < \ldots < \tau_n = N$ and the real part

$$R\{\varepsilon^{-h} \rho_{\tau_j}(t, \varepsilon)\} > R\{\varepsilon^{-h} \rho_{\tau_{j+1}}(t, \varepsilon)\}; \quad j = 1, \ldots, n-1. \tag{46}$$

Denote a truncated formal matrix solution by

$$T(t, \sigma) = \mathfrak{J}(t, \sigma)E(t, \varepsilon) \text{ where } \mathfrak{J}(t, \sigma) = \sum_{j=0}^{m} \sigma^j \tilde{X}_j(t), \tag{47}$$

and m is sufficiently large, i.e. $m \geq 3p + H$. See (45) for the significance of p. Replace equation (1) by

$$\sigma^H \dot{\bar{X}}(t, \sigma) = B(t, \sigma) \bar{X}(t, \sigma) \tag{48}$$

where $\bar{X}$ is a square matrix; $H = h\, r$; and $B(t, \varepsilon^{1/r}) \equiv A(t, \varepsilon)$. Likewise rewrite the asymptotic expansion for A so that

$$B(t, \sigma) \sim \tilde{B}(t, \sigma) = \sum_{i=0}^{\infty} \sigma^i B_i(t)$$

where $B_{nr}(t) \equiv A_n(t)$, $(n = 0, 1, 2, \ldots)$, and all other B_i's are identically zero. The symbol $\sim$, as used here, will indicate throughout this paper an asymptotic expansion valid to an infinite number of terms in the sense of Poincaré(1).

Then replace (48) by the equivalent equation

$$\sigma^H \dot{\bar{X}} - C \bar{X} = D \bar{X} \tag{49}$$

where

$$D(t, \sigma) = B(t, \sigma) - C(t, \sigma)$$

and

$$C(t, \sigma) = \sigma^H \dot{T}(t, \sigma)\, T^{-1}(t, \sigma), \tag{50}$$

noting that the truncated solution T satisfies the equation $\sigma^H \dot{T} = C\, T$. Equation (49) is treated as a non-homogeneous equation with the right member known; and hence its general solution

$$\bar{X}(t, \sigma) = T(t, \sigma)K(\sigma) + \sigma^{-H}T(t, \sigma) \int_{\gamma_{ij}}^{t} T^{-1}(\tau, \sigma)D(\tau, \sigma)\bar{X}(\tau, \sigma)d\tau, \tag{51}$$

where matrix $K(\sigma)$ is composed of elements which may vary with σ, but not with t. The indicated integral is a matrix of the form

$$\left\| \int_{\gamma_{ij}}^{t} q_{ij}(t, \sigma)dt \right\|.$$

The N^2 different lower limits of integration γ_{ij} need not all have the same value. In fact to keep certain exponentials which appear later on uniformly bounded, take

(52) $$\gamma_{ij} = a \text{ if } j \leq i \text{ and } \gamma_{ij} = b \text{ if } j > i; \ (i, j = 1, \ldots, N).$$

Integral equation (51) is not only equivalent to the original differential equation (48); but also it can and will be used to show that, when $K(\sigma) = I$, T is an asymptotic approximation to the corresponding solution $\bar{X}$.

Observe that formally $\tilde{B}(t, \sigma) = \sigma^H \dot{\tilde{X}}(t, \varepsilon) \tilde{X}^{-1}(t, \varepsilon)$ and on comparing this with (50) and remembering the series $\mathfrak{X}$ and $\mathfrak{Y}$ agree up through the mth power of σ, it is evident the asymptotic expansions of B and C agree at least up to the power σ^{m+1-p}; i.e.

$$D(t, \sigma) = B(t, \sigma) - C(t, \sigma) = \sigma^{m+1-p}F(t, \sigma)$$

where the elements $f_{ij}(t, \sigma)$ of $F(t, \sigma)$ satisfy the inequalities

(53) $$|f_{ij}(t, \sigma)| < M_m, \text{ a constant}; \ (i, j = 1, \ldots, N)$$

in the domain D_m; i.e. provided t is on $[a, b]$, ε is in R, and $0 < |\varepsilon| \leq \varepsilon_m \leq \varepsilon_1$. In D_m the elements of $T(t, \sigma)$, $T^{-1}(t, \sigma)$, $C(t, \sigma)$, abd $D(t, \sigma)$ are continuous in t and analytic in σ.

Setting $\bar{X}(t, \sigma) = U(t, \sigma)T(t, \sigma)$, (51) becomes

$$U = T K T^{-1} + \sigma^{m+1-p-H} T\left(\int_{\gamma_{ij}}^{t} T^{-1} F U T \, d\tau \right) T^{-1}$$

To estimate the size of the elements $\Psi_{ij}(t, \sigma)$ in the matrix

$$\Psi(t, \sigma) = T(t, \sigma)\left(\int_{\gamma_{ij}}^{t} T^{-1} F U T \, d\tau \right) T^{-1}(t, \sigma),$$

let $L(\sigma)$ be the largest numerical maximum attained by the elements of $U(t, \sigma)$ as t varies over $[a, b]$. In view of (47)

$$T^{-1} = E^{-1} \mathfrak{T}^{-1}.$$

Let the elements of $\mathfrak{T}$ be $\tau_{ij}(t, \sigma)$ and those of $\mathfrak{T}^{-1}$ be $\hat{\tau}_{ij}(t, \sigma)$. Then there exists a positive constant k_m such that in D_m

(54) $$|\tau_{ij}(t, \sigma)| < k_m \text{ and } |\hat{\tau}_{ij}(t, \sigma)| < k_m/|\sigma|^p; \ (i, j = 1, \ldots, N).$$

Hence utilizing the bounds in (53) and (54), and recalling (46) and the special values chosen for the lower limits of integration γ_{ij} in (52), it is clear that

(55) $$|\Psi_{ij}(t, \sigma)| < N^5 k_m^4 M_m(b-a) L(\sigma)/|\sigma|^{2p}.$$

Consider next the particular solution of (48), say $\bar{X}_0(t, \sigma)$, which satisfies the initial conditions

$$\bar{X}_0(a, \sigma) = T(a, \sigma).$$

Since $T(a, \sigma)$ is analytic in σ, so is $\bar{X}_o(t, \sigma)$. Moreover every solution of (48) can be written in the form

$$\bar{X}(t, \sigma) = \bar{X}_o(t, \sigma)\mathcal{H}(\sigma)$$

where the elements of matrix $\mathcal{H}$ may vary with σ, but not with t. Substituting this expression for $\bar{X}$ into (51) and solving for K and setting $t = a$

$$K(\sigma) = T^{-1}(a, \sigma)\,\{\bar{X}_o(a, \sigma) - \sigma^{m+1-p-H}T(a, \sigma)\int_{\gamma_{1j}}^{a} T^{-1}\,F\,\bar{X}_o d\tau\}\,\mathcal{H}(\sigma);$$

i.e.

$$K(\sigma) = \mathcal{P}(\sigma)\,\mathcal{H}(\sigma) \tag{56}$$

where the elements of matrix $\mathcal{P}(\sigma)$ are analytic in σ.

To show that the determinant $|\mathcal{P}(\sigma)|$ does not vanish for sufficiently small values of σ; suppose that $|\mathcal{P}(\sigma_o)| = 0$. Then if we take $K(\sigma_o) \equiv 0$, there will exist an $\mathcal{H}_1(\sigma_o)$ satisfying (56) such that $\mathcal{H}_1(\sigma_o) \not\equiv 0$. Since $\bar{X}_o(t, \sigma_o) \not\equiv 0$; there is a corresponding solution $\bar{X}_1(t, \sigma_o) = \bar{X}_o(t, \sigma_o)\mathcal{H}_1(\sigma_o)$ such that $\bar{X}_1(t, \sigma_o) \not\equiv 0$. Also

$$\bar{X}_1(t, \sigma_o) = \sigma_o^{m+1-p-H}\,T(t, \sigma_o)\int_{\gamma_{1j}}^{t} T^{-1}(\tau, \sigma_o)F(\tau, \sigma_o)\bar{X}_1(\tau, \sigma_o)d\tau$$

and with

$$\bar{X}_1(t, \sigma_o) = U_1(t, \sigma_o)\,T(t, \sigma_o),$$

$$U_1(t,\sigma_o) = \sigma_o^{m+1-p-H}T(t,\sigma_o)\,\{\int_{\gamma_{1j}}^{t} T^{-1}(\tau,\sigma_o)F(\tau,\sigma_o)U_1(\tau,\sigma_o)T(\tau,\sigma_o)d\tau\}\,T^{-1}(t,\sigma_o). \tag{57}$$

As in (55) the absolute value of any element in the matrix in the right member of equation (57) is less than $N^5k_m^4L(\sigma_o)\,|\sigma_o|^{m+1-3p-H}\,M_m(b-a)$ and the absolute value of some element in the left member attains the maximum value $L(\sigma_o)$ for some particular value of t on $[a, b]$. Therefore for this element

$$L(\sigma_o) < N^5\,k_m^4\,L(\sigma_o)\,|\sigma_o|^{m+1-3p-H}\,M_m(b-a)$$

but this is impossible if $|\sigma_o|$ is sufficiently small. To avoid this contradiction it must follow that the determinant $|\mathcal{P}(\sigma)|$ does not vanish when $|\sigma|$ is sufficiently small, say when $0 < |\varepsilon| \leq \varepsilon_{2,m} \leq \varepsilon_m$.

This means that if we take $K(\sigma) = I$ there is a corresponding unique matrix $\mathcal{H}_2(\sigma)$ analytic in σ which satisfies the equation $I = \mathcal{P}(\sigma)\mathcal{H}_2(\sigma)$. This $\mathcal{H}_2(\sigma) \not\equiv 0$ and the solution $\bar{X}_2(t, \sigma) = \bar{X}_o(t, \sigma)\mathcal{H}_2(\sigma)$ is also not identically zero. Furthermore with $\bar{X}_2 = U_2T$

$$U_2 = I + \sigma^{m+1-p-H}\,T(\int_{\gamma_{1j}}^{t} T^{-1}F\,U_2T\,d\tau)\,T^{-1}.$$

Reasoning as before

$$L(\sigma) < 1 + N^5 k_m^4 L(\sigma) |\sigma|^{m+1-3p-H} M_m(b-a)$$

and

$$L(\sigma)[1 - N^5 k_m^4 |\sigma|^{m+1-3p-H} M_m(b-a)] < 1.$$

Hence for sufficiently small σ's, say those for which

$$0 < |\varepsilon| \leq \varepsilon_{3,m} \leq \varepsilon_{2,m},$$

the $L(\sigma) < 2$. Consequently

$$U_2(t, \sigma) = I + \sigma^{m+1-3p-H} \mathcal{B}_m(t, \sigma).$$

In this equality and in subsequent equations in this paper the symbol $\mathcal{B}_m$ indicates a matrix whose elements in absolute value are all uniformly bounded, the bound depending on m, but not on σ or t, in the domain under consideration. The precise value of the elements of $\mathcal{B}_m$ in one equation may differ from those in the next equation. Also $\bar{X}_2 = [\mathcal{J} + \sigma^{m+1-3p} \mathcal{B}_m \mathcal{J}]E$ or $\bar{X}_2(t, \sigma) = [\mathcal{J}(t, \sigma) + \sigma^{m+1-3p} \mathcal{B}_m(t, \sigma)]E(t, \varepsilon)$. These findings are summarized in the following

THEOREM 2. If the hypotheses i-vi in §1 and vii in §9 are satisfied, then to each integer $m > 3p$ there correspond N independent vector solutions of (1) which can be assembled into a square matrix $\bar{X}_m(t, \varepsilon)$ such that

$$\bar{X}_m(t, \varepsilon) = [\sum_{j=0}^{m} \varepsilon^{j/r} \tilde{X}_j(t) + \varepsilon^{(m+1)/r} \mathcal{B}_m(t, \varepsilon)]E(t, \varepsilon) \tag{58}$$

where each element in $\mathcal{B}_m$ in absolute value is less than $\mathcal{K}_m$, a constant, when t is in $[a, b]$, ε is in R, and $0 < |\varepsilon| \leq \varepsilon_m \leq \varepsilon_1$.

In short the truncated formal solutions do represent solutions of (1) asymptotically to a finite number of terms; but Theorem 2 falls short in two respects from our final objective. Asymptotic representation of a given solution was wanted which would be valid for any value of m in a fixed region; in the theorem both the solution and the region vary with m.

§10. CERTAIN ASYMPTOTIC SOLUTIONS

In this section an attempt will be made to free our results from their dependence on the cut-off point m.

Set $t = a$ in the formal matrix solution to get the series

$\tilde{X}(a, \epsilon) = \sum_{k=0}^{\infty} \sigma^k \tilde{X}_k(a)$ running in powers of σ. Then construct a matrix

(59) $$G(\sigma) \sim \sum_{k=0}^{\infty} \sigma^k \tilde{X}_k(a)$$

when ϵ is in R and $0 < |\epsilon| \leq \epsilon_1$ with elements $g_{ij}(\sigma)$ which are analytic in σ in the domain under consideration, (for details see Tamarkin and Besikowitsch[21]).

Next define a set of solutions of (48), say $\bar{X}_3$, by the initial conditions

$$\bar{X}_3(a, \sigma) = G(\sigma).$$

Since $\bar{X}_m(t, \epsilon)$ represents N independent vector solutions of (1), there exists a matrix $\mathcal{L}(\epsilon)$ with elements analytic in ϵ and constant with respect to t such that

(60) $$\bar{X}_3(t, \sigma) = \bar{X}_m(t, \epsilon)[I + \mathcal{L}(\epsilon)],$$

when t is in [a, b]; ϵ is in R; and $0 < |\epsilon| \leq \epsilon_m$. At $t = a$

$$G(\sigma) = \bar{X}_m(a, \epsilon) + \bar{X}_m(a, \epsilon)\, \mathcal{L}(\epsilon).$$

Solving for $\mathcal{L}(\epsilon)$,

$$\mathcal{L}(\epsilon) = \bar{X}_m^{-1}(a, \epsilon)[G(\sigma) - \bar{X}_m(a, \epsilon)].$$

Writing $||d_{ij}(\sigma)|| = G(\sigma) - \bar{X}_m(a, \epsilon)$, it is evident in view of (58) and (59) that the absolute values $|d_{ij}(\sigma)| < |\sigma|^{m+1} M_m$, where M_m is a constant; ϵ is in R, and $0 < |\epsilon| \leq \epsilon_m$. Then because of (45) and (58) it follows that the absolute values of the elements in $\mathcal{L}(\epsilon)$ are less than $|\sigma|^{m+1-p} M_{2,m}$, where $M_{2,m}$ is an appropriate constant, ϵ is in R, and $0 < |\epsilon| \leq \epsilon_{2,m} \leq \epsilon_m$.

Also rewrite (60) as

$$\bar{X}_3 = [T + \sigma^{m+1} \mathcal{B}_m + (T + \sigma^{m+1} \mathcal{B}_m)E\, \mathcal{L} E^{-1}]\, E.$$

By virtue of the bounds on the elements of $\mathcal{B}_m$ and $\mathcal{L}$ and (46) it follows that

$$\bar{X}_3(t, \sigma) = [\sum_{k=0}^{m-p} \sigma^k \tilde{X}_k(t) + \sigma^{m-p+1} \hat{\mathcal{B}}_m(t, \sigma)]\, E(t, \epsilon)$$

where the absolute values of the elements in the first r_1 columns of $\hat{\mathcal{B}}_m$ are less than an appropriate constant $M_{3,m}$ when t is on [a, b], ϵ is in R, and $0 < |\epsilon| \leq \epsilon_{3,m} \leq \epsilon_{2,m}$. This domain of variation for ϵ can be extended at once to $0 < |\epsilon| \leq \epsilon_1$ by increasing $M_{3,m}$ if necessary, for in the portion of R where $\epsilon_{3,m} \leq |\epsilon| \leq \epsilon_1$ the elements of

$$(\bar{X}_3 E^{-1} - \sum_{k=0}^{m-p} \sigma^k \tilde{X}_k)/\sigma^{m-p+1}$$

are uniformly bounded. This follows from the fact that $A(t, \varepsilon)$ is continuous in t and analytic in ε; the initial values of $G(\sigma)$ are analytic in ε and therefore the solution $\bar{X}_3(t, \sigma)$ is continuous in t and analytic in ε in the domain under consideration. Note particularly that these results are valid for all $m > p$ and that $\bar{X}_3$ is independent of m. These facts are summarized in

> THEOREM 3. If the hypothesis i-vii are valid, then the first τ_1 columns of the formal solution $\tilde{X}$ are asymptotic solutions of equation (1) in the domain D_1.

This result still falls short of the final objective whenever $\tau_1 < N$. In order to extend the results to all columns we plan to utilize the τ_1 known vector solutions; reduce the original to one of rank $N - \tau_1$; reapply Theorem 3 to the new system and get $(\tau_2 - \tau_1)$ more solutions, etc.; finally arriving at a full set of N independent solutions with the desired asymptotic expansions.

For this purpose consider a vector equation

$$\varepsilon^h \dot{Y} = B\,Y \tag{61}$$

and the corresponding matrix equation

$$\varepsilon^h \dot{\bar{Y}} = B\,\bar{Y}\,. \tag{62}$$

Let

$$Y = \begin{Vmatrix} Y_{11} & Y_{12} \\ Y_{21} & Y_{22} \end{Vmatrix} \quad \text{and} \quad B = \begin{Vmatrix} B_{11} & B_{12} \\ B_{21} & B_{22} \end{Vmatrix} \tag{63}$$

where Y_{11} and B_{11} are both square matrices of order τ_1. Assume that τ_1 vector solutions

$$\begin{Vmatrix} Y_{11} \\ Y_{21} \end{Vmatrix}$$

of (61) are known and that <u>the determinant</u> $|Y_{11}| \neq 0$.

Then these solutions can be used to reduce the order of (62). To do so define matrix S by the equation $S = Y_{21}Y_{11}^{-1}$ and make the transformation

$$\begin{Vmatrix} Y_{11} & Y_{12} \\ Y_{21} & Y_{22} \end{Vmatrix} = \begin{Vmatrix} I & 0 \\ S & I \end{Vmatrix} \begin{Vmatrix} Z_{11} & Z_{12} \\ Z_{21} & Z_{22} \end{Vmatrix}; \tag{64}$$

whereupon (62) takes the form $\varepsilon^h \dot{\bar{Z}} = C\,\bar{Z}$ where

$$\bar{Z} = \begin{Vmatrix} Z_{11} & Z_{12} \\ Z_{21} & Z_{22} \end{Vmatrix} ; \qquad C = \begin{Vmatrix} C_{11} & C_{12} \\ 0 & C_{22} \end{Vmatrix} ;$$

(65) $$C_{11} = B_{11} + B_{12}S; \quad C_{12} = B_{12}; \quad \text{and} \quad C_{22} = B_{22} - S\,B_{12}$$

Thus equation (62) can be decomposed into two separate systems,

(66) $$\varepsilon^h \dot{Z}_{22} = C_{22}Z_{22}$$

and

(67) $$\varepsilon^h \dot{Z}_{12} = C_{11}Z_{12} + C_{12}Z_{22} ;$$

the first is of order $(N - \tau_1)$ and the second of order τ_1.

The known matrix Y_{11} satisfies the equation

$$\varepsilon^h \dot{Y}_{11} = B_{11}Y_{11} + B_{12}Y_{21} = (B_{11} + B_{12}S)\,Y_{11} = C_{11}Y_{11} .$$

Hence, if a fundamental matrix solution Z_{22} is known for (66) the general solution Z_{12} of (67) can be found by quadratures, for

(68) $$Z_{12} = Y_{11}K + \varepsilon^{-h}\,Y_{11} \int_b^t Y_{11}^{-1}\,C_{12}Z_{22}d\tau$$

where K is a matrix which may vary with ε, but not with t. The corresponding solutions of (62) are $Y_{12} = Z_{12}$ and $Y_{22} = Z_{22} - S\,Z_{12}$.

There is one obstacle to employing such a reduction to equation (1). The matrix $X_{11}(t, \varepsilon)$, which corresponds to Y_{11} in the reduction, may vanish on $[a, b]$, thus voiding the procedure at least for the entire interval $[a, b]$. To overcome this obstacle systems (1) will be transformed into a more amenable equation as specified by the following theorem.

§11. A CANONICAL FORM FOR DIFFERENTIAL EQUATION (1)

THEOREM 4. If the hypotheses i-vi in §1 are satisfied, there exists a substitution

(69) $$X = H(t, \varepsilon)\,Y,$$

where

(70) $$H(t, \varepsilon) = \sum_{k=0}^{m} \varepsilon^{k/r}\,H_k(t)$$

and r and m are suitable positive integers, which will transform equation (1) into a canonical equation

(71) $$\varepsilon^h \dot{Y} = B(t, \varepsilon)\,Y$$

where

(72) $$B(t, \varepsilon) \sim \tilde{B}(t, \varepsilon) = \| \delta_{ij}\rho_j(t, \varepsilon) \| + \varepsilon^h \sum_{k=0}^{\infty} \varepsilon^{k/r}\,B_k(t).$$

This asymptotic expansion is valid when t is in $[a, b]$, ε is in R, and $0 < |\varepsilon| \leq \varepsilon_2 \leq \varepsilon_1$. Moreover in this same domain all the elements $b_{ij}(t, \varepsilon)$ of $B(t, \varepsilon)$ possess derivatives of all orders with respect to t and are analytic in ε; all the elements $h_{ijk}(t)$ of $H_k(t)$ and all the elements $b_{ijk}(t)$ of $B_k(t)$, $(k = 0, 1, \ldots)$, also possess derivatives of all orders in t on $[a, b]$. In the expansion of the determinant

$$|H(t, \varepsilon)| = \varepsilon^{p/r} \sum_{k=0}^{\infty} \varepsilon^{k/r} h_k(t)$$

the p is a suitable non-negative integer and the lead term $h_0(t) \neq 0$ on $[a, b]$. The $\rho_j(t, \varepsilon)$'s in (72) are the same as those in (44).

PROOF. As a first step in the proof, we intend to show that <u>proceeding formally</u> there is a substitution

$$X = H_f Y \quad \text{where} \quad H_f(t, \varepsilon) = \sum_{k=0}^{\infty} \sigma^k H_k(t) \tag{73}$$

which will reduce (1) to the form

$$\varepsilon^h \dot{Y} = B_f(t, \varepsilon) Y \tag{74}$$

where

$$B_f = \begin{Vmatrix} M_1 & 0 & \cdots & 0 \\ 0 & M_2 & & \vdots \\ \vdots & & \ddots & \vdots \\ 0 & \cdots & \cdots & M_n \end{Vmatrix} \tag{75}$$

and where each square matrix M_i has the form

$$M_i = \rho_i(t, \varepsilon) I + \varepsilon^h \sum_{j=0}^{\infty} \sigma^j M_{ij}(t), \quad i = 1, \ldots, n. \tag{76}$$

This fact is verified by checking back through the seven cases considered in §2-8. In cases I and II equation (1) is in the desired form (74) at the outset. In case III a series of normalizing transformations of type (11) and exponential transformations of type (16) reduce (1) to an equation of type (6) where $h = 0$. If the successive transformations are incorporated into a single substitution of type (73) and the exponential transformations are all omitted it is clear that (1) is transformed into the desired form (74) with $n = 1$.

In case IV a sequence of normalizing and exponential transformations combined with one zero-inducing transformation (21) reduces (1) essentially to N distinct systems each of the first order. Therefore if the exponential substitutions are omitted and the others used and combined into a single transformation (73), (1) is reduced to (74) with $n = N$.

In case V a sequence of normalizing and exponential transformations combined with one zero-inducing transformation essentially splits (1) into K separate systems (24) of lower order. If the exponential transformations are omitted and the other substitutions incorporated into a single substitution $X = P(t, \varepsilon)Z$, this substitution will transform (1) into an equation of type (22), equivalent to K separate systems

$$\varepsilon^h \dot{Z}_j = G_j Z_j, \quad j = 1, \ldots, K, \tag{77}$$

each of order lower than N, see (24).

If at this stage there exist transformations

$$Z_j = H_{fj}(t, \varepsilon) Y_j, \quad j = 1, \ldots, K,$$

where each $H_{fj}(t, \varepsilon)$ has the same form as $H_f(t, \varepsilon)$ in (73) such that (77) becomes

$$\varepsilon^h \dot{Y}_j = B_{fj}(t, \varepsilon) Y_j, \quad j = 1, \ldots, K,$$

where each B_{fj} has the desired structure indicated in (75) and (76); then the transformation

$$X = P(t, \varepsilon) = \left\| \begin{matrix} H_{f1}(t, \varepsilon) & 0 & 0 \\ 0 & H_{f2}(t, \varepsilon) & \vdots \\ \vdots & & \ddots \\ 0 & & H_{fK}(t, \varepsilon) \end{matrix} \right\| Y$$

of type (73) will reduce (1) as desired to form (74).

This observation permits one to confine attention to a single equation of type (77). Again a zero-inducing substitution (21) is used on (77) to produce a new system

$$\varepsilon^h \dot{Z} = \sum_{k=0}^{\infty} \varepsilon^k C_k Z \tag{78}$$

where the C's have the structure indicated in (26) and (28). At this stage two alternatives occur: [I] A finite sequence of exponential, shearing, normalizing, and zero-inducing transformations reduces equation (78) to one falling under case I. Omission of the exponential transformations, but use of all the others, will result in this case in a reduction to the desired form (74); or [II] the system is split into several systems. If the exponential transformations are omitted, the same splitting occurs; and the process is continued until eventually in a finite number of steps (1) is reduced to a form (74). Hence the desired single formal transformation of type (73) exists.

Since only non-singular transformations are involved, in the formal expansion of the determinant

$$|H_f(t, \varepsilon)| = \varepsilon^{p/r} \sum_{k=0}^{\infty} \varepsilon^{k/r} h_k(t)$$

the p is either zero or a suitable positive integer and the lead term $h_o(t) \neq 0$ on $[a, b]$.

Once substitution (73) is known we have merely to use the first $(m + 1)$-terms in this expansion to get substitution (70) mentioned in Theorem 4, provided $m \geq rh + p$. The statements in Theorem 4 relating to differentiability are readily verified; hence Theorem 4 is proved.

§12. A FUNDAMENTAL SYSTEM OF ASYMPTOTIC SOLUTIONS

In this section a canonical system

$$\varepsilon^h \dot{Y} = B Y \tag{79}$$

such as (71) is considered. Since ρ_{τ_i}, ρ_{τ_j}, and the difference $(\rho_{\tau_i} - \rho_{\tau_j})$ are all polynomials in σ, we may define the rational numbers h_{ij} by writing

$$\rho_{\tau_i}(t, \varepsilon) - \rho_{\tau_j}(t, \varepsilon) \equiv \sigma^{(h-h_{ij})r} \rho_{ij}(t) + \text{higher powers of } \sigma,$$

$(i, j = 1, \ldots, n)$, where $\rho_{ij}(t) \neq 0$ on $[a, b]$ and $i \neq j$. Note that if $i \neq j$, $h_{ij} = h_{ji}$, and $1/r \leq h_{ij} \leq h_j$; $i, j = 1, \ldots, n$. Set $h_{ii} = 0$. By direct substitution it can be verified that the canonical system (79) possesses N independent formal vector solutions which can be assembled into a matrix of the form $\tilde{Y} = F E$ where

$$F = \begin{Vmatrix} F_{11} & F_{12} & \cdots & F_{1n} \\ F_{21} & F_{22} & & \\ \vdots & & \ddots & \\ F_{n1} & & & F_{nn} \end{Vmatrix}; \quad E = \begin{Vmatrix} E_1 & 0 & \cdots & 0 \\ 0 & E_2 & & \\ \vdots & & \ddots & \\ 0 & & & E_n \end{Vmatrix};$$

$$F_{ij} = \varepsilon^{h_{ij}} \sum_{k=0}^{\infty} \sigma^k F_{ijk}(t); \qquad i, j = 1, \ldots, n,$$

$$E_i = I \exp \{q_{\tau_i}(t, \varepsilon)\} ; \qquad i = 1, \ldots, n;$$

and the orders of the square matrices F_{ii} and E_i are alike and equal to $(\tau_i - \tau_{i-1})$. Note particularly that the determinants $|F_{ii0}(t)| \neq 0$ on $[a, b]$; $i = 1, \ldots, n$.

THEOREM 5. *The canonical system*

$$\varepsilon^h \dot{Y} = B Y \tag{80}$$

possesses N independent vector solutions which can be assembled into a matrix $\bar{Y}$ such that $\bar{Y}(t, \varepsilon) \sim \tilde{Y}(t, \varepsilon)$ when t is in $[a, b]$, ε is in R, and $0 < |\varepsilon| \leq \varepsilon_2 \leq \varepsilon_1$.

PROOF. This theorem will be proved by induction on n, the number of distinct ρ_j's. As shown in §10, if $n = 1$, Theorem 5 is certainly valid. Suppose therefore it is also valid whenever the number of distinct ρ_j's does not exceed $n - 1$. To complete the induction begin by replacing (80) by the matrix equation $\varepsilon^h \dot{\bar{Y}} = B \bar{Y}$ and split the matrices $\bar{Y}$ and B into four blocks as shown in (63). Also split the known formal expansions $\tilde{Y}$ and $\tilde{B}$ into four blocks; thus

$$\tilde{Y} = \begin{Vmatrix} \tilde{Y}_{11} & \tilde{Y}_{12} \\ \tilde{Y}_{21} & \tilde{Y}_{22} \end{Vmatrix} \quad \text{and} \quad \tilde{B} = \begin{Vmatrix} \tilde{B}_{11} & \tilde{B}_{12} \\ \tilde{B}_{21} & \tilde{B}_{22} \end{Vmatrix} .$$

By hypothesis Y_{11} and Y_{21} are both known, as well as their asymptotic expansions, i.e.

$$\begin{Vmatrix} Y_{11} \\ Y_{21} \end{Vmatrix} \sim \begin{Vmatrix} \tilde{Y}_{11} \\ \tilde{Y}_{21} \end{Vmatrix}$$

in the domain under consideration. In the same domain $B \sim \tilde{B}$. Likewise split F and E; thus

$$(81) \qquad F = \begin{Vmatrix} \Psi_{11} & \Psi_{12} \\ \Psi_{21} & \Psi_{22} \end{Vmatrix} = \sum_{k=0}^{\infty} \sigma^k \begin{Vmatrix} \Psi_{11k}(t) & \Psi_{12k}(t) \\ \Psi_{21k}(t) & \Psi_{22k}(t) \end{Vmatrix} ; \quad E = \begin{Vmatrix} E_{11} & 0 \\ 0 & E_{22} \end{Vmatrix} ,$$

where $\Psi_{11} = F_{11}$ and $E_{11} = E_1$. In particular

$$\tilde{Y}_{11} = \Psi_{11}E_{11}; \ \tilde{Y}_{12} = \Psi_{12}E_{22}; \ \tilde{Y}_{21} = \Psi_{21}E_{11}; \ \text{and} \ \tilde{Y}_{22} = \Psi_{22}E_{22} .$$

Since $Y_{11} \sim \Psi_{11} \exp \{q_{\tau_1}\}$ and the determinant $|F_{110}(t)| \neq 0$, the determinant $|Y_{11}(t, \varepsilon)| \neq 0$ when t is in $[a, b]$, ε is in R, and $0 < |\varepsilon| \leq |\varepsilon_3| \leq |_1$. The reduction to an equation of order $(N - \tau_1)$ therefore can and will be carried out as described in the last portion of §10.

To do so note first, that since $S = Y_{21}Y_{11}^{-1}$,

$$S \sim \tilde{S} = \begin{Vmatrix} F_{21} \\ \cdot \\ \cdot \\ \cdot \\ F_{n1} \end{Vmatrix} F_{11}^{-1} = \sum_{k=0}^{\infty} \sigma^k S_k(t).$$

Then referring to (65) and (72) and letting

$$B_k(t) = \begin{Vmatrix} B_{11k} & B_{12k} \\ B_{21k} & B_{22k} \end{Vmatrix} ,$$

it is evident each C has an asymptotic expansion; specifically

$$C_{11} \sim \tilde{C}_{11} = \rho_1 I + \varepsilon^h \sum_{k=0}^{\infty} \sigma^k (B_{11k} + B_{12k}\tilde{S}) ;$$

$$C_{12} \sim \tilde{C}_{12} = \varepsilon^h \sum_{k=0}^{\infty} \sigma^k B_{12k} ;$$

and

$$C_{22} \sim \tilde{C}_{22} = ||\delta_{ij}\rho_j(t, \varepsilon)|| + \varepsilon^h \sum_{k=0}^{\infty} \sigma^k(B_{22k} - \tilde{S} B_{12k})$$

where $i, j = \tau_1 + 1, \tau_1 + 2, \ldots, N$.

Hence equation

(82) $$\varepsilon^h \dot{Z}_{22} = C_{22} Z_{22}$$

has the structure prescribed in Theorem 5 and the asymptotic expansion for C_{22} contains only $(n - 1)$ distinct ρ_j's. Hence by hypothesis equation (82) possesses $(N - \tau_1)$ actual vector solutions Z_{22} asymptotic respectively to the $(N - \tau_1)$ independent formal vector solutions $\tilde{Z}_{22}$, i.e. $Z_{22} \sim \tilde{Z}_{22}$.

To compute $\tilde{Z}_{22}$ make the formal transformation

$$\begin{Vmatrix} \tilde{Y}_{11} & \tilde{Y}_{12} \\ \tilde{Y}_{21} & \tilde{Y}_{22} \end{Vmatrix} = \begin{Vmatrix} I & 0 \\ \tilde{S} & I \end{Vmatrix} \begin{Vmatrix} \tilde{Z}_{11} & \tilde{Z}_{12} \\ \tilde{Z}_{21} & \tilde{Z}_{22} \end{Vmatrix} ,$$

the analogue of (64). As a consequence

$$Z_{22} \sim \tilde{Z}_{22} = \tilde{Y}_{22} - S \tilde{Y}_{12} = (\mathcal{Y}_{22} - \tilde{S} \mathcal{Y}_{12}) E_{22} .$$

Also $\tilde{Z}_{12} = \tilde{Y}_{12} = \mathcal{Y}_{12} E_{22}$.

Turn next to the equation

$$\varepsilon^h \dot{Z}_{12} = C_{11} Z_{12} + C_{12} Z_{22}$$

and its general solution

$$Z_{12} = Y_{11}K + \varepsilon^{-h} Y_{11} \int_b^t Y_{11}^{-1} C_{12} Z_{22} \, d\tau.$$

Set

(83) $$W = \varepsilon^h \frac{d}{dt} (Y_{11}^{-1} Z_{12}) = Y_{11}^{-1} C_{12} Z_{22}$$

and introduce the new symbols $\Psi_k(t)$ and $\Phi_k(t)$ defined by the expansions

$$\mathcal{Y}_{11}^{-1} (\sum_{k=0}^{\infty} \sigma^k B_{12k})(\mathcal{Y}_{22} - \tilde{S} \mathcal{Y}_{12}) = \sum_{k=0}^{\infty} \sigma^k \Psi_k(t)$$

and

$$\mathcal{Y}_{11}^{-1} \mathcal{Y}_{12} = \sum_{k=0}^{\infty} \sigma^k \Phi_k(t).$$

With this symbolism

$$W \sim \varepsilon^h E_{11}^{-1} \sum_{k=0}^{\infty} \sigma^k \Psi_k(t) E_{22}$$

or

$$W = \varepsilon^h E_{11}^{-1} \left\{ \sum_{k=0}^{m} \sigma^k \Psi_k(t) + \sigma^{m+1} \mathcal{B}_m(t, \varepsilon) \right\} E_{22} .$$

Formally (83) is equivalent to

$$\varepsilon^h \frac{d}{dt} [E_{11}^{-1} \sum_{k=0}^{\infty} \sigma^k \Phi_k(t) E_{22}] = \varepsilon^h E_{11}^{-1} \sum_{k=0}^{\infty} \sigma^k \Psi_k(t) E_{22}$$

or

$$\sum_{k=0}^{\infty} \sigma^k \Psi_k = \sum_{k=0}^{\infty} \sigma^k \dot{\Phi}_k + (\sum_{k=0}^{\infty} \sigma^k \Phi_k) \| \varepsilon^{-h}(\rho_i - \rho_1) \delta_{ij} \|$$

where $i, j = \tau_1 + 1, \tau_2 + 2, \ldots, N$. Hence, returning to actual rather than formal equations,

$$\varepsilon^h \frac{d}{dt} [E_{11}^{-1} (\sum_{k=0}^{m+hr} \sigma^k \Phi_k) E_{22}] = \varepsilon^h E_{11}^{-1} [\sum_{k=0}^{m} \sigma^k \Psi_k + \sigma^{m+1} \mathcal{B}_m] E_{22} .$$

Consequently

$$(84) \quad \varepsilon^{-h} \int_b^t W \, dt = [E_{11}^{-1} (\sum_{k=0}^{m+hr} \sigma^k \Phi_k) E_{22}] \Big|_b^t + \sigma^{m+1} \int_b^t E_{11}^{-1} \mathcal{B}_m E_{22} d\tau.$$

Again following Tamarkin and Besikowitsch(21) construct a matrix

$$\mathfrak{F}(\sigma) \sim \sum_{k=0}^{\infty} \sigma^k \Phi_k(b)$$

with elements analytic in σ in the domain under consideration. In (68) set

$$K = K(\sigma) = E_{11}^{-1} (b, \varepsilon) \mathfrak{F}(\sigma) E_{22}(b, \varepsilon);$$

use (84), and observe that

$$\mathfrak{F}(\sigma) = \sum_{k=0}^{m+hr} \sigma^k \Phi_k(b) + \sigma^{m+hr+1} \mathcal{B}_m(\sigma) ;$$

whereupon (88) takes the form

$$(85) \quad Z_{12} = \{\sigma^{m+hr+1} Y_{11}(t, \varepsilon) E_{11}^{-1}(b, \varepsilon) \mathcal{B}_m(\sigma) E_{22}(b, \varepsilon) E_{22}^{-1}(t, \varepsilon) + Y_{11}(t, \varepsilon) E_{11}^{-1}(t, \varepsilon) [\sum_{k=0}^{m+hr} \sigma^k \Phi_k(t)] + \sigma^{m+1} Y_{11}(t, \varepsilon) [\int_b^t E_{11}^{-1} \mathcal{B}_m E_{22} d\tau] E_{22}^{-1}(t, \varepsilon) \} E_{22}(t, \varepsilon).$$

Since $Y_{11} \sim \sum_{k=0}^{\infty} \sigma^k y_{11k} E_{11}$, see (81),

$$Y_{11} = \sum_{k=0}^{m+hr} \sigma^k y_{11k} E_{11} + \sigma^{m+hr+1} \mathcal{B}_m E_{11} .$$

Substituting this expression for Y_{11} in (85) we find that

(86) $Z_{12} = \{\sigma^{m+1} B_m(t, \sigma) + (\sum_{k=0}^{m+hr} \sigma^k \mathcal{Y}_{11k}(t))(\sum_{k=0}^{m+hr} \sigma^k \Phi_k(t))\}E_{22}(t, \varepsilon)$,

for the elements in the matrices

$$P_1 = E_{11}(t, \varepsilon)E_{11}^{-1}(b, \varepsilon)B_m(\sigma)\, E_{22}(b, \varepsilon)E_{22}^{-1}(t, \varepsilon)$$

and

$$P_2 = E_{11}(t, \varepsilon)(\int_b^t E_{11}^{-1} B_m E_{22} d\tau)E_{22}^{-1}(t, \varepsilon)$$

are uniformly bounded by virtue of (46).

If we write $\mathcal{Y}_{11}^{-1} = \sum_{k=0}^{\infty} \sigma^k \hat{\mathcal{Y}}_{11k}$; then formally

$$(\sum_{k=0}^{\infty} \sigma^k \mathcal{Y}_{11k})(\sum_{k=0}^{\infty} \sigma^k \hat{\mathcal{Y}}_{11k}) = I .$$

Also formally

$$\sum_{k=0}^{\infty} \sigma^k \Phi_k = (\sum_{k=0}^{\infty} \sigma^k \hat{\mathcal{Y}}_{11k})(\sum_{k=0}^{\infty} \sigma^k \mathcal{Y}_{12k}).$$

Hence actually

$$(\sum_{k=0}^{m+hr} \sigma^k \mathcal{Y}_{11k})(\sum_{k=0}^{m+hr} \sigma^k \Phi_k) = \sum_{k=0}^{m+hr} \sigma^k \mathcal{Y}_{12k} + \sigma^{m+hr+1} B_m$$

and (86) becomes

$$Z_{12} = \{\sigma^{m+1} B_m(t, \sigma) + \sum_{k=0}^{m} \sigma^k \mathcal{Y}_{12k}\}E_{22}(t, \varepsilon).$$

Since this equality is valid for any large value of m and Z_{12} is defined independently of m, $Z_{12} \sim \mathcal{Y}_{12}E_{22}$.

Since $Z_{22} \sim (Y_{22} - \tilde{S}\mathcal{Y}_{12})E_{22}$; $S \sim \tilde{S}$; $Y_{12} = Z_{12}$; and $Y_{22} = Z_{22} + S Z_{12}$; it follows that in the domain under consideration

$$Y_{12} \sim \mathcal{Y}_{12} E_{22} = \tilde{\mathcal{Y}}_{12}$$

and

$$Y_{22} \sim \mathcal{Y}_{22} E_{22} = \tilde{\mathcal{Y}}_{22}. \quad \text{Q. E. D.}$$

Because (69) is a non-singular transformation, we have also proved

THEOREM 6. If the hypotheses i-vii are satisfied, there exists a positive number $\varepsilon_2 > 0$ such that the formal solutions (43) of equation (1) are asymptotic solutions when t is in $[a, b]$, ε is in R, and $0 < |\varepsilon| \leq \varepsilon_2 \leq \varepsilon_1$.

§13. THE NON-HOMOGENEOUS EQUATION

Again consider a matrix equation $\varepsilon^h \dot{\overline{V}} = A \overline{V}$ of type (1) subject to hypotheses i-vii and add to the right member a matrix $C(t, \varepsilon)$ to get the non-homogeneous equation

(87) $$\varepsilon^h \dot{\overline{V}} = A \overline{V} + C$$

where

(88) $$C(t, \varepsilon) \sim \tilde{C} = \sum_{k=0}^{\infty} \sigma^k C_k(t)\, G(t, \sigma);$$

$\sigma = \varepsilon^{1/r}$; r is a suitable integer;

$$G(t, \sigma) = \|\delta_{ij} \exp \{ \int_a^t \varepsilon^{-h} g_j(t, \varepsilon) dt \}\| ;$$

and

$$g_j(t, \varepsilon) = \sum_{k=0}^{hr-1} \sigma^k g_{jk}(t); \quad i, j = 1, \ldots, N.$$

HYPOTHESES:

If t is in $[a, b]$, ε is in R, and $0 < |\varepsilon| \leq \varepsilon_2$, then

(viii) The asymptotic expansion (88) is valid.

(ix) All elements $c_{ij}(t, \varepsilon)$ of the square matrix $C(t, \varepsilon)$ possess derivatives of all orders with respect to t and are analytic in ε.

(x) If $c_{ijk}(t)$ is the element in the ith row and j^{th} column of $C_k(t)$, $(i, j = 1, \ldots, N;\ k = 0, 1, 2, \ldots)$, each $c_{ijk}(t)$ possesses derivatives of all orders on $[a, b]$.

(xi) Either the real part

$$R\{\varepsilon^{-h}[g_j(t, \varepsilon) - \rho_i(t, \varepsilon)]\} \geq 0 \quad \text{or} \quad \leq 0$$

for each pair of values (i,j); $i, j = 1, \ldots, N$.

(xii) The $g_{jk}(t)$ possess derivatives of all orders on $[a, b]$.

(xiii) For each pair of integers (i,j), $(i, j = 1, \ldots, N)$, either

(89) $$g_j(t, \varepsilon) - \rho_i(t, \varepsilon) \equiv 0$$

or

(90) $$g_j(t, \varepsilon) - \rho_i(t, \varepsilon) = \sigma^{H_{ij}} p_{ij}(t) + \text{higher powers of } \sigma,$$

where $p_{ij}(t) \neq 0$ on $[a, b]$.

Equation (90) serves to define certain of the integers H_{ij} where $H_{ij} = H_{ji}$ and $0 \leq H_{ij} \leq rh-1$. If (89) applies, set $H_{ij} = H_{ji} = rh$. Subject to these hypotheses a particular solution of (87) will be specified and its asymptotic expansion computed. The general solution of (87) is

$$\bar{V} = \bar{X}\, K(\sigma) + \varepsilon^{-h} \bar{X} \int_{\Gamma_{ij}}^{t} \bar{X}^{-1}\, C\, d\tau \tag{91}$$

where the matrix $K(\sigma)$ is temporarily unspecified. The lower limit of integration corresponding to the elements in the ith row and j^{th} column in the product $(\bar{X}^{-1} C)$ is Γ_{ij} as indicated. Set $\Gamma_{ij} = a$ if $R\{\varepsilon^{-h}[g_j - \rho_i]\} \geq 0$ or is identically zero; and set $\Gamma_{ij} = b$ if $R\{\varepsilon^{-h}[g_j - \rho_i]\} \leq 0$. The matrix $\bar{X}$ in (91) is the fundamental solution of the equation $\varepsilon^h \dot{\bar{X}} = A\, \bar{X}$ with the asymptotic expansion $\bar{X} \sim \tilde{X} = \mathcal{X} E$, where

$$\mathcal{X} = \sum_{k=0}^{\infty} \sigma^k \tilde{X}_k ; \quad \mathcal{X}^{-1} = \sigma^{-p} \sum_{k=0}^{\infty} \sigma^k \hat{X}_k ;$$

and

$$E = \| \delta_{ij} \{ \int_a^t \varepsilon^{-h} \rho_j \, dt \} \| .$$

First we wish to evaluate $\tilde{U}$ formally if

$$\tilde{U} = \varepsilon^{-h} \int^t \tilde{X}^{-1} \tilde{C} d\tau \quad \text{and} \quad \dot{\tilde{U}} = \sigma^{-p-rh} E^{-1} \left(\sum_{k=0}^{\infty} \sigma^k \Phi_k \right) G \tag{92}$$

where the Φ's are defined by the equation

$$\sum_{k=0}^{\infty} \sigma^k \Phi_k = \left(\sum_{k=0}^{\infty} \sigma^k \hat{X}_k \right) \left(\sum_{k=0}^{\infty} \sigma^k C_k \right).$$

Let $\tilde{U} = \sigma^{-p} E^{-1} \left(\sum_{k=0}^{\infty} \sigma^k \| \sigma^{-H_{ij}} \Psi_{ijk} \| \right) G$; and then

$$\dot{\tilde{U}} = \sigma^{-p} E^{-1} \{ \| -\delta_{ij}\, \varepsilon^{-h} \rho_j \| \left(\sum_{k=0}^{\infty} \sigma^k \| \sigma^{-h_{ij}} \Psi_{ijk} \| \right) +$$

$$\sum_{k=0}^{\infty} \sigma^k \| \sigma^{-H_{ij}} \dot{\Psi}_{ijk} \| + \left(\sum_{k=0}^{\infty} \sigma^k \| \sigma^{-H_{ij}} \Psi_{ijk} \| \right) \| \delta_{ij}\, \varepsilon^{-h} g_j \| \} \, G.$$

Comparing this result with (92),

$$\sum_{k=0}^{\infty} \sigma^k \Phi_k = \sum_{k=0}^{\infty} \sigma^k \| \sigma^{-H_{ij}} (g_j - \rho_i) \Psi_{ijk} \| + \sigma^{rh} \sum_{k=0}^{\infty} \sigma^k \| \sigma^{-h_{ij}} \dot{\Psi}_{ijk} \| .$$

Then by equating the coefficients of successive powers of σ in (93) the Ψ_{ijk}'s can be computed since the Φ_k's are known. For example if $g_j - \rho_i = 0$, $H_{ij} = rh$ and

$$\Psi_{ijk} = \int_a^t \varphi_{ijk}\, d\tau .$$

With the Ψ's evaluated as indicated we come to our final

THEOREM 7. If the hypotheses in i-xiii are satisfied, there exists a particular solution of (87), say $\overline{V}_p$, such that $\overline{V}_p \sim \tilde{X}\,\tilde{U}$ when t is in $[a, b]$, ε is in R, and $0 < |\varepsilon| \leq \varepsilon_3 \leq \varepsilon_2$.

PROOF. Since $\overline{X}^{-1} = \sigma^{-p} E^{-1} \left\{ \sum_{k=0}^{m} \sigma^k \hat{X}_k + \sigma^{m+1} \mathcal{B}_m \right\}$ and

$$C = \left\{ \sum_{k=0}^{m} \sigma^k C_k + \sigma^{m+1} \mathcal{B}_m \right\} G;$$

$$\sigma^{-rh}\, \overline{X}^{-1} C = \sigma^{-p-rh} E^{-1} \left(\sum_{k=0}^{m} \sigma^k \Phi_k + \sigma^{m+1} \mathcal{B}_m \right) G$$

$$= \sigma^{-p} \frac{d}{dt} \left\{ E^{-1} \left(\sum_{k=0}^{m} \sigma^k || \sigma^{-H_{ij}} \Psi_{ijk} || \right) G \right\} + \sigma^{-p-rh+m+1} E^{-1} \mathcal{B}_m G.$$

Therefore

$$(94)\quad \varepsilon^{-h} \int_{\Gamma_{ij}}^{t} \overline{X}^{-1} C\, d\tau = \left\{ E^{-1} \left(\sum_{k=0}^{m} \sigma^k || \sigma^{-H_{ij}} \Psi_{ijk} || \right) G \right\} \Big|_{\Gamma_{ij}}^{t} + \sigma^{-p-rh+m+1} \int_{\Gamma_{ij}}^{t} E^{-1} \mathcal{B}_m G\, d\tau.$$

Introducing the functions $f_{ij}(\sigma)$, analytic in σ, such that

$$f_{ij}(\sigma) \sim \sum_{k=0}^{\infty} \sigma^{k-H_{ij}} \Psi_{ijk}(\Gamma_{ij}, \sigma),$$

see reference (21), and then define $\overline{V}_p$ by the equation

$$\overline{V}_p = \overline{X} K_p + \varepsilon^{-h} \overline{X} \int_{\Gamma_{ij}}^{t} \overline{X}^{-1} C\, d\tau$$

where $K_p = ||K_{ij}||$ and $K_{ij} = f_{ij}(\sigma) \exp \left\{ \int_a^{\Gamma_{ij}} \varepsilon^{-h}(g_j - \rho_i) dt \right\}$.

Utilizing this special value of K_p and (94) in (95) it can be shown that $\overline{V}_p \sim \tilde{X}\,\tilde{U}$ by proceeding in a fashion entirely analogous to that given in §12. In this way Theorem 6 is proved, although for brevity these final details will be omitted.

BIBLIOGRAPHY

(1) POINCARÉ, H., "Sur les intégrales irrégulières des équations linéaires," Acta Math. 8(1886) 295-344.

(2) LIOUVILLE, J., "Sur le développement des fonctions ou parties de fonctions en séries dont les divers termes sont assujettes à satisfaire à une même équation différentielle du second ordre contenant un paramètre variable," Liouville Jour. 2 (1837) 16-35.

(3) HORN, J., "Über eine lineare Differentialgleichung zweiter Ordnung mit einem willkürlichen Parameter," Math. Annalen 52(1899) 271-292.

(4) HORN, J., "Über lineare Differentialgleichungen mit einem veränderlichen Parameter," Math. Annalen 52(1899) 340-362.

(5) SCHLESINGER, L., "Über asymptotische Darstellungen der Lösungen linearer Differentialsysteme als Funktionen eines Parameters," Math. Annalen 63(1907) 277-300.

(6) BIRKHOFF, G. D., "On the asymptotic character of the solutions of certain linear differential equations containing a parameter," Trans Amer. Math. Soc. 9 (1908) 219-231.

(7) NOAILLON, P., "Développements asymptotiques dans les equations différentielles linéaires à paramètre variable," Memoires de la Soc. des Sci. de Liège, 3rd Ser., 11 (1912) 197 pp.

(8) TAMARKIN, J. D., "Some general problems of the theory of ordinary linear differential equations and expansions of an arbitrary function in series of fundamental functions," Math. Zeitschrift 27 (1927) 1-54; also thesis (Russian) Petrograd (1917).

(9) PERRON, O., "Über die Abhängigkeit der Integrale eines Systems linearer Differentialgleichungen von einem Parameter," Sitzungsberichte der Heidelberger Akad. der Wissenschaften," Math.-Naturw., Abh. 13(1918), Abh. 15 (1918), und Abh. 3 (1919).

(10) BIRKHOFF, G.D., LANGER, R. E., "The boundary problems and developments associated with a system of ordinary differential equations of the first order," Proc. Amer. Acad. of Arts and Sci., No. 2, 58 (1923) 51-128.

(11) TRJITZINSKY, W. J., "Theory of linear differential equations containing a parameter," Acta Math. 67 (1936) 1-50.

(12) TURRITTIN, H. L., "Asymptotic solutions of certain ordinary differential equations associated with multiple roots of the characteristic equation," Amer. Jour. of Math. 58 (1936) 364-376.

(13) GRADSTEIN, I. S., "Linear equations with variable coefficients and small parameters in the highest derivatives," (Russian) Mat. Sbornik N. S. 27(69), (1950) 47-68.

(14) JEFFREYS, H., JEFFREYS, B. S., "Methods of Mathematical Physics," Cambridge Univ. Press (1946) 491-495.

(15) LANGER, R. E., "The asymptotic solutions of ordinary linear differential equations of the second orders, with special reference to the Stokes phenomena," Bull. Amer. Math. Soc. 40 (1934) 545-582.

(16) LANGER, R. E., "The asymptotic solutions of ordinary linear differential equations of the second order, with special reference to a turning point," Trans. Amer. Math. Soc. 67 (1949) 461-490.

(17) CHERRY, T. M., "Uniform asymptotic formulae for functions with transition points," Trans. Amer. Math. Soc. 68 (1950) 224-257.

(18) EVANS, R. L., "Asymptotic solutions in the neighborhood of a turning point for linear ordinary differential equations containing a parameter," thesis, Univ. of Minn. April 1951.

(19) DICKSON, L. E., Modern Algebraic Theories, 1930, p. 105.

(20) TRJITZINSKY, W. J., "Singular point problems in the theory of linear differential equations," Bull. Amer. Math. Soc. 44 (1938) 209-223.

(21) TAMARKIN, J., BESIKOWITSCH, A., "Über die asymptotischen Ausdrücke für die Integrale eines Systems linearer Differentialgleichungen, die von einem Parameter abhängen," Math. Zeitschrift, 21 (1924) 119-125.

(22) WASOW, W., "On the construction of periodic solutions of singular perturbation problems," Annals of Math. Studies, No. 20, pp. 313-350, Princeton Univ. Press, 1950.

NOTE: This paper was prepared under contract with the Office of Naval Research, and was equally sponsored by the Office of Air Research.

Ingram Content Group UK Ltd.
Milton Keynes UK
UKHW051522290523
422428UK00005B/12

9 780691 095813